Universe is Universal

LESSONS FROM PLANETS TO THE STARS

MOHIT JAIN

BLUEROSE PUBLISHERS
India | U.K.

For permissions requests or inquiries regarding this publication,
please contact:

BLUEROSE PUBLISHERS
www.BlueRoseONE.com
info@bluerosepublishers.com
+91 8882 898 898
+4407342408967

ISBN: 978-93-5989-078-4

Cover Design: Muskan Sachdeva
Typesetting: Pooja Sharma

First Edition: October 2024

Contents

Chapter 1
The Earth

Q.1 What is Earth?

Ans- Earth is our home planet. Scientists believe Earth and its moon formed around the same time as the rest of the solar system. They think that was about 4.5 billion years ago. Earth is the fifth largest Planet in the Solar system. Its diameter is about 8,000 miles. And Earth is the third-closest Planet to the sum. Its average distance from the sun is about 93 million miles. only Mercury and Venus are closer.

Earth has been called the "Goldilocks Planet". In the story of "Goldilocks and the three Bears" a little girl named Goldilocks liked everything just right. Her porridge couldn't be too hot or too cold. And her bed couldn't be too hard or too soft. On Earth, everything is just right for life to exist. It's warm, but not too warm and it has water, but not too much water.

Earth is the only Planet known to have large amount of liquid water. Liquid water is essential for life. Earth is the only Planet where life is known to exist.

Q.2 What Does Earth Look Like?

Ans- From Space, Earth looks like blue marble with white swirls and areas of brown, yellow, green and white. The blue is water, which covers about 71 Percent of Earth's surface. The white swirls are clouds. The areas of brown, yellow, and green are land. And the areas of white are ice and snow.

The equator is an imaginary circle that divides Earth into two halves. The northern half is called the Northern Hemisphere. The southern half is called the southern Hemisphere. The northernmost point on earth is called the North Pole. The southernmost point on Earth is called the South Pole.

Q.3 How do we know Earth is Round?

Ans- Humans have known that Earth is round for more than 2000 years! The ancient Greeks measured shadows during summer solstice and also calculated Earth's circumference. They used positions of stars and constellations to estimate distances on Earth. They could even see the planet's round shadow on the moon during a lunar eclipse. (We still can see this during lunar eclipses.)

Today, Scientists we geodesy, which in the science of measuring Earth's shape, gravity and rotation. Geodesy provides accurate measurements that show Earth it round with GPS and other satellites, scientists can measure Earth's size and shape to within a centimeter. Pictures from space also show Earth is round like Moon. Even through our Planet is a sphere, it is not a perfect sphere. Because of the force caused when Earth rotates, the North and south Poles are slightly flat. Earth's Rotation, Wobbly motion and other forces are making the Planet change state very slowly, but it is still round.

Q.4 How Does Earth Move ?

Ans- Earth orbits, the sun once every 365 days, or one year. The shape of its orbit is not quite a Perfect Circle. It's more like an oval, which causes Earth's distance from the sun to vary during the year. Earth is nearest the Sun, or at perihelion, "in January where it's about 11 million miles away. Earth is farthest from the sun, or at "aphelion, in July when it's about 95 million miles away.

At the equator, Earth spins at Just over 1,000 miles Per hour. Earth makes a full spin around its axis once every 24 hours, or one day, the axis is an imaginary line through the center of the Planet from the North Pole to the south Pole. Rather than straight and down, Earth's axis is tilted at an angle of 23.5 degrees.

Q.5 Why do we have Day and Night?

Ans- At all times, Half of Earth is lighted by the sum and half is in darkness. Areas facing toward the Sun experience day time. Area facing away from the sun experience night time. As the Planet spins most places on Earth Cycle through day and night once every 24 hours. The North Pole and South Pole have continuous day light in darkness depending on the time of year.

Q.6 Why Does Earth Have Seasons?

Ans- Earth has seasons because its axis is tilted. Thus, the sun's rays hit different parts of the Planet more directly depending on the time of year.

From June to August, the sun's rays hit the Northern Hemisphere more directly than the Southern Hemisphere. The result is warm (summer) weather in the Northern Hemisphere and cold (winter) weather in the Southern Hemisphere. From December to February, the sun's ray hit the Northern Hemisphere less directly than the Southern Hemisphere. The result is cold (winter) weather in the Northern Hemisphere and warm (summer) weather in the Southern Hemisphere. From September to November the sun shines equally on both hemispheres. The result is fall in the Northern Hemisphere and spring in the Southern Hemisphere. The sum also shines equally on both hemispheres from March to May. The result is spring in the Northern Hemisphere and fall in the Southern Hemisphere.

Q.7 What are Earth's Different Parts?

Ans- Earth consists of land, air, water and life. The land contains mountains, valleys and flat areas. The air is made up of different gases, mainly nitrogen and oxygen. The water includes oceans, lakes, rivers, streams, rain, snow and ice. Life consists of people, animals and Plants. There are millions of species or kinds of life, on Earth. Their sizes range from very time to very large.

Below Earth's surface are layers of rock and metal. Temperatures increase with depth, all the way to about 12000 degrees Fahrenheit at Faith's inner core.

Earth's parts once were seen at largely separate from each other. But now they are viewed together as the "Earth System". Each part connects to and affects each of the other Parts. For example-

- Clouds in the air drop rain and snow on land.
- Water gives life to Plants and animals
- Volcanoes on land send gas and dust into the air
- People breathe air and drink water.

Earth System science is the study of interactions between and among Earth's different Parts.

Q.8 Why and how does NASA Study Earth?

Ans- NASA studies Earth to learn about how the planet changes. Earth's Parts - land, air, water and life are always changing. Some of the changes are natural and some are caused by Humans. Scientists want to understand how Earth has changed in the past and how it is changing now. This information helps them predict how Earth might change in the future.

NASA studies Earth using satellites. Satellites look toward Earth from Space. They take pictures of and collect information about all Earth Parts. NASA satellites are especially good for observing clouds, oceans, land, and ice. They also measure gases in the atmosphere, such as ozone and Carbon dioxide. They measure how much energy enters and leaves Earth's atmosphere. And they monitor wildfires, Volcanoes and their smoke.

Information gathered by NASA satellites helps scientists predict weather and climate. It also helps public health officials track disease and famine. It helps farmers decide when to plant crops and what kinds to plant. And it helps emergency workers respond to natural disasters. The more people know about Earth and its current and predicted changes, the better decisions they can make.

Chapter 2
Plant Growth and Developed

Plants required to grow is Air, water, soil and Sun Light.

Green plants convert solar energy to chemical energy by the process of photosynthesis. This reaction is directly or indirectly responsible for nearly all life on earth. During Photosynthesis, carbon dioxide, a gas is combined with water and a solar energy converted to carbohydrates, a solid formation of Carbohydrates is a chemical way to store the sun's energy as food. Carbohydrates produced from Photosynthesis provide energy for all plant growth and maintenance.

Q.1 Have you ever wondered how a small seed turns into a new plant?

Ans- We will discuss the whole process of how plants, grow in great detail. We will also study the parts of the plants.

A plant is a living thing that grows in the ground after the seed is sowed. It has a lot of leaves coming out of the stem and branches. A plant also has flowers and fruits in it which are very useful in nature. The plants also has roots through which they get water and other minerals from the ground. A plant contributes to nature in a lot of ways. It gives oxygen to all living organism. A Plants keeps the environment clean and beautiful. Plants Provides fruits, vegetables, flowers and green leaves to humans and animals. Plants also make their own food through the Photosynthesis process. Plants can also be grown in water bodies such as Lakes, Rivers, Ponds, Seas and Oceans. Plants are an important part of the ecosystem as well.

Q.2 What is a seed?

Ans- A seed is sown inside the ground and it the embryo of the plant. The Process of seed development is also called germination. Through the process of germination, the embryo sprouts with the help of water and soil and grow into small stem roots, and leaves. The process leads to the sprouting of the embryo which is the first step toward the growth of plant.

Q.3 How do plants grow?

Ans- After the process of germination, the embryo will start sprouting into small leaves, roots and a stem the roots will then start spreading in the ground which leads to the growth of the Plant. The roots absorbs water and minerals and then the stem starts growing. With the help of sun light, the stem grows in the branches, green leaves start growing out of the branches. The Five things plants need to grow are sun light, water, minerals, Air, and soil for preparing food. With the passage of time fruits and flowers start growing and the plant grows fully.

The process of photosynthesis. The Process through which plants makes their own food is called photosynthesis. The plants use sunlight to make food. They takes carbon dioxide and release oxygen into the air. All humans and animals use oxygen for breathing purposes while humans do the opposite. This whole process is called the process of photosynthesis. The sum is a renewable energy source that plays a pivotal role in our everyday life, from warming the earth to the water cycle it is an essential part of our daily existence.

Chapter 3
The Water Cycle on Earth

Water is essential to Life on Earth. In its three Phases (solid, Liquid and Gas). Water ties together the major parts of the earth's climatic system - air, clouds the ocean, Lakes vegetation snow pack and Glaciers.

The water cycle shows the continuous movement of water within the earth and atmosphere. It is a complex system that includes many different processes. Liquid water evaporates into water vapaur condenses to form clouds and precipitates back to Earth in the form of rain and snow. Water is different phases moves through the atmosphere (transportation) liquid water flows across land (run off) into the ground (infiltration and percolation) and through the ground (ground water) ground water moves into plants (plant uptake) and evaporated from plants into the atmosphere (transpiration). Solid ice and snow can turn directly in the gas (sublimation). The opposite can also take place when water vapour becomes solid.

Water, society and ecology.

Water influences the intensity of the climate variability and change. It is the key part of extern events. Such as drought and floods. It's abundance and timely delivery are critical for meeting the needs of society and ecosystems. Humans or we can say all living organisms including plants use water for drinking, industrial applications, irrigating Agriculture, hydropower, waste disposal and recreation. It is important that water sources are protected both for human uses and ecosystem health. In many areas water supplies are being depleted because of population growth, pollution and development.

Chapter 4
Air

Q.1 What is Air?

Ans- Air is all around us, but we can't see it. So what is air, exactly? It's a mixture of different gases. The air is Earth's atmosphere is made up of approximately 78% Percent nitrogen and 21% percent oxygen. Air also has small amount of other gases, too such as carbon dioxide, neon and hydrogen.

Air is the Earth, atmosphere. Air is a mixture of many gases and tiny dust particles. It is the clear gas in which living things live and breath. It has an indefinite shape and volume. It has mass and weight because it is matter. The weight of air creates atmospheric pressure. There is no air in outer space. Air is need to Animals live and need to breath the oxygen in the air. In breathing the lungs, put oxygen into the blood and send back carbon dioxide in the air to live. They give off the oxygen that we breath. without it we die.

Air can be polluted by some gases (such as carbon monoxide, hydrocarbons and nitrogen oxides smoke and ash. The air pollution cause various problem including smog, acid rain and global warming. It can damage people's health and the environment. There are debates about weather or not act climate change, but soon enough the earth will hear up too much, causing our home to become too hot and not support life! Some say less people would die of cold weather and that is there true but there is already a huge amount of people dying from heart and that number is and will keep increasing at a freighting height. Since early times air has been used to create technology. Ships moved with sails and wind mills used the mechanical motion of air. Air crafts use propellers to move air over a wing, which allows them to fly pneumatics use air pressure to move things. Since the late 1900's, air power is also used to generate electricity.

Chapter 5
Soil

An estimated 70% percent of Earth's surface is covered with water, while the remaining 30% percent constituted land. The layer of Earth that is composed of soil and is influenced by the process of soil formation is called Pedosphere. But what exactly is soil and what is soil made up of? Technically, the soil is a mixture that contains minerals, organic matter and living organisms. But broadly speaking, soil can refer to any loose sediment. Moreover there are many type of soil that are distributed around the world and there are generally classified into the following

1- Clay soil (2-) sandy soil (3)- Loamy soil (4) silt soil typically, the soil consists of 45% mineral 50% empty spaces or voids and 5% organic matter. Furthermore, soil performs many important functions such as

1) Providing a growth medium for the Plants
2) 0
3) One of the most crucial components of the biosphere
4) Provides habitat for organism

Q.1 How is soil formed?

Ans- Soil is formed by weathering of rocks solid rock can weather away in one of the three ways into the soil namely.

1- **Mechanical weathering-** This is commonly observed near the surface of the earth also called physical weathering, as this process is influenced by physical forces such as wind, water and temperature.

2- **Chemical weathering**- As the name suggests chemical weathering occurs when rocks are broken down by chemical reactions often such types of weathering can change the chemical components of the soil.

3- **Biological weathering**- Through not an actual weathering process living organisms weaken and subsequently disintegrate rocks often by initiating mechanical or chemical weathering. For instance tree roots can grow into cracks in the rock prying them apart and causing mechanical fractures, Microorganisms can secrete chemicals that can increase the rocks susceptibility to weathering.

4- **Composition of soil**- The soil is composed of different components- 5% organic matter, 45% minerals 20-30% different gases and 20-30% water. Therefore the soil is known as a heterogeneous body. Given the composition of the soil

5- **Organic Matter**- Organic substances is found in very small amounts in the soil. Plants and animals are the main source of organic matter. Depending upon the decomposition stage, the organic matter is of the following three type

A- Completely decomposed organic matter.

B- Partially decomposed organic matter.

C- Undecomposed organic matter.

6- **Minerals**- Minerals are an important element of the soil. These are solid components composed of the atoms. These occur naturally and have a fixed chemical composition olivine and Feld spar the main minerals present in the soil.

7- **Gaseous components**- The air filled pores of the soil contain the gaseous components. Nitrogen and oxygen present in the pores is generally the atmospheric air fixed by the microorganisms. However the composition of the carbon dioxide is higher due to the gas produced by microorganisms present in the soil.

8- **Water**- The soil dissolves the minerals and nutrions in the water and transports it to different parts of the plants. These are essential for the growth and development of the plant importance of soil.

Soil is an important element essential for the survival of the living organisms. The importance of soil is mentioned below.

1) The fertile soil helps in the growth and development of the plants. The plants thus produced are healthy and provide food, clothing, furniture and medicines.

2) It supports many life forms including bacteria's, fungi, algae etc. These microbes in turn maintain environment balance by retaining the moisture and decaying the dead organisms.

3) The top soil supports certain life activities such as reproduction, hatching, nesting breeding etc. of a few organisms.

4) The organic matters present in the soil increase the fertility of the soil which is responsible for the growth of the plants. It also contains certain minerals and elements that are necessary for the plants to carry out their cellular activities.

5) Soil is use for making cups, utensils, tiles etc. The contents in the soil such as gravel, clay and sand are used in the construction of homes, rodes and buildings etc.

6) Useful minerals medicines such as calcium, iron and other substances such as petroleum jelly for cosmetic are extracted from the soil.

7) The soil absorbs the rain water. This water is evaporated and released in the air during sunny days, making the atmosphere cooler.

Chapter 6
Sun Light

The electromagnetic radiation of the sun that reach the earth, particularly infrared, visible light and ultra violet light.

Sunlight refers to the electromagnetic radiation of the sun that reaches the earth, particularly infrared, visible light and ultraviolet light. The electromagnetic radiation consist of electromagnetic waves that can be characterized by frequency or wavelength of oscillations. However, not all of it reaches the earth. The earth's atmosphere filters the electromagnetic radiation allowing a part of it to pass through especially visible light. Most of the sun radiation does not reach the earth's surface. Some of it is absorbed in the atmosphere while others are scattered to the space.

In terms of photosynthesis the most essential part of the sun's radiation is the visible light. The visible light is the part of the electromagnetic radiation with wave lengths ranging between 380 nm and 750 nm. The length of the wavelength is inversely proportional to the amount of energy. It means that the shorter the wavelength, the higher the energy there is accordingly red and blue of the visible light are the one rest absorbed by the plant pigment, chlorophyll and also the ones most useful as energy for the light reactions of Photosynthesis.

The sun is the star at the center of the solar system. It is nearly perfect ball of hot plasma heated to incandescence by nuclear fusion reaction its cake. The Sun radiates this energy mainly light ultraviolet and infrared radiation and is the most importance source of energy for life on Earth. Sun, star around which earth and the other components of the solar system revolve. It is the dominant body of the system, constituting more than 99% percent of its entire mass. The sun is the source of an enormous amount of energy, a portion of which provides earth with the light and heat necessary to support life on this universe. Light or visible light is electromagnetic radiation that can be perceived by the human eye and animals eye visible light is usually defined as having wavelengths in the range of 400-700nm nenometers. Corresponding to frequencies of 150-420 terahertz between the infrared (with longer wavelengths and the ultraviolet. In physics the term light may refer more broadly to electromagnetic radiation of any wave length, whether visible or not. In this sense gamma rays, X-rays, microwaves and radio waves the also light. The primary properties of light are intensity propagation direction frequency or wave length spectrum and polarization. It's speed in a vacuum, 299792458 meters a second (m/s) it one of the fundamental constants of nature. Like all types of electromagnetic radiation, visible light propagates by massless elementary particles called photons that represents the quanta of electromagnetic field and can be analyzed as both waves and particles. The study of light known as optics is an important research area in modern physics. The main Source of natural light on this universe of Earth is the sun historically, another important source of light for humans has been fire, from ancient campfire to modern kerosene lamps. With the development of electric lights and power system electric lighting has effectively replaced fire lights.

Chapter 7
Thunder Storms

Thunder storms and electricity have a very strong relationship. We can discuss this relationship in discovery of electricity. First we understand Thunder storms. Thunder storms include thick clouds heavy rain fall or hail, lightning, thunder and strong winds. Thunder storms happens when hot moist air rises quickly to cooler parts of the atmosphere. There the air cools and clouds and rain form. Lightning which is a form of electricity, develops inside the clouds. As the lightning treats the air, it causes it to expand. This causes the sound of thunder meanwhile, cooled air sinks towards the ground. This movement of air causes heavy wind's.

Thunder and Lightning,

Thunder and lightning happen at the same time. Lightning is a quick powerful burst of electricity. The burst can be up to five miles long. Lightning is born inside a storm cloud. It can stay inside a cloud or it can flash horizontally to nearby cloud. It can also shoot from a cloud toward the ground below. Lightning is extremely fast. It is also very hot. An average bolt of lightning travel more than 220,000 miles per hour. It can be 53000°F. That is more than five times the heat of the sun. Thunder is the sound caused by a bolt of lightning. When lightning strikes, the intensity of the flash causes the air around it to rapidly expand. The sound of this expanding air is thunder. Light travel faster than sound. So lightning is often seen before thunder is heard after an initial crack or boom, thunder may seem to roll for a few seconds. This sound is actually the full length of lightning's shock waves. The blast of air closest to the Earth's surface is heard first. It is followed by the shock wave's more-distant rumbles.

Chapter 8
Electricity

Q.1 Who discovered The Electricity?

Ans- Electricity is a form of energy that exist in nature since electricity occurs in nature it was not invented it can be discovered by seeing thunder storms humans got an idea. And thinking about this lightning is torn inside a storm cloud. Thunder storms includes thick clouds heavy rainfall, lightning. Thunder and lightning happen at the same time lightning is a quick powerful burst of electricity. So let us brand out who discovered the electricity?

Electricity was discovered and understood by many scientists. But Benjamin Franklin it give the credit for discovering electricity in the year 1752. Benjamin Franklin conducted an experiment using a kite and key on a rainy day. He wanted to demonstrate the relationship between lightning and electricity. He flew the kite tied with the key during the thunderstorms. As he expected, the electricity from the storm clouds transferred to the key and he got a shock. By feeling that shock be after understands that relationship between lightning and electricity is same as he expected the electricity is travels them the storm clouds. After the relationship is developed then the question is how to use this?

After the revelations of Franklin's experiment many scientists started building upon his work and started studying and understanding the concept of electricity. Scientists have found evidence that ancient people had experimented with electricity much before Franklin's discovery of electricity. In 600 B.C., the ancient Greeks found Static electricity by rubbing fur on amber. In the year 1600, English Physician William Gilbert used the word electricity to explain the force between the objects when they are rubbed with each other.

The first practical application of electricity was the invention of electric dynamo by Michal Faraday in the year 1831. This set an electrical revolution around the world. This universe is got lightning by Thomas Edison invention he invented the light bulb in the year 1878. In the late 1800's, Nikola Tesla invented alternating current and induction motor.

Discovery of Electricity

Do you rely on electricity like you to food, water and air? What would like be like without electricity to power our homo appliances and even the light? Electricity is a universal power which is used by everyone in many ways in daily life. Without this we can't live. Just thinks without electricity. We wouldn't be able to enjoy our daily wonder of the day what a horrible thought! But don't worry electricity does exist and it allows us to enjoy life in so many ways since electricity is a natural force that exists in our world, it didn't have to be invented. I did, however have to be discovered and understood. Most people give credit to Benjamin Franklin for discovering electricity by his most famous experiment using a kite and key on a rainy day. Benjamin Franklin had one of the greatest scientific minds of his time he was an American scientist. He was interested in many areas of science made many discoveries and invented many things including bifocal glasses. In the mid-1700's, he become interested in electricity. Up until that time, scientists had mainly known about and experiment with static electricity Benjamin Franklin took thinks a big step head. He come up with the idea that electricity had Positive and Negative elements and that electricity flowed between these elements. He also believed that lightning was a form of this flowing electricity. In 1752, Franklin conducted his famous kite experiment. In order to show that lighting with electricity, he flew during a thunder storm. He tied a metal key to the kite string to conduct the electricity. Just as he thought, electricity from the storm clouds transferred to the kite and electricity flowed down the string and gave him a shock then touching the key. He's lucky that he didn't get hurt, but he didn't mind the shock since it provide his idea.

Building your Franklin's walk many other scientists studied electricity and began to understand more about how it work. For eq. - In 1879 Thomas Edison patented the electric light bulb and our world has been brighter ever since! This is a universally used in every where.

But was Benjamin Franklin really the first person to discover electricity? May be not! At the turn of the 17th century English scientist William Gilbert established the science underlying the study of electricity and magnetism. Inspired by Gilbert's work, another English man, Sir Thomas brown made further investigation and wrote books about his findings. Gilbert and Browne are credited with being the first scientists to use the term electricity scientists have found evidence that ancient people may have experimented with electricity too. In 1936, a Clay Pot was discovered that suggested that the first batteries may have been invented over 2000 years ago. The Clay Pot contained copper Plates, tin alloy, and an iron rod. It could have been used to create an electricity current by filling it with an acidic solution like vinegar. No one knows that the device was used for, but it sheds some light on the fact that people may have been learning about electricity long before Benjamin Franklin.

Q. But who Discovered Electricity?

Ans- Electricity is a form of energy and it occurs in nature, this was discovered by Benjamin Franklin, so it was not invented. As to who discovered it, many misconceptions abound. Some give credit to Benjamin Franklin for discovering electricity but his experiments only helped establish the connection between lightning and electricity, nothing more. The trouth about the discovery of electricity is a bit more complex than a man flying his kite. It actual goes back make then two thousand year.

In about 600 B.C. the Ancient Greeks discovered that rubbing fur on amber (Fossilized tree resin) caused an attraction between two and so what the Greeks discovered was actually static electricity. Additionally, researchers and archeologists in the 1930 discovered Pot with sheets of copper inside that they believe may have been ancient batteries meant to produced light at ancient Roman sites similar devices were found in archeological digs near Baghdad meaning ancient Persians may have also used an early form of batteries But by the 17th century many electricity related discoveries had been made, such as the invention of an early electrostatic generator the differentiation between positive and negative currents and the classification of materials as conductors of insulators. In the year 1600, English Physician William Gilbert used the Latin word "electrics" to describe the force that certain substances exert when rubbed against each other. A few year later another English scientist Thomas Browne wrote several books and he used the word electricity to describe his investigations based on Gilberts work. In 1752 Ben Franklin experiment already proved that simply proved that lightning and tiny electric sparks were the same thing.

Italian Physicist Alessando Volta discovered that particular chemical reactions could produce electricity, and in 1800 he constructed the voltaic Pile (an early electric battery) that produced a steady electric current, and so he was the first person to create a steady flow of electric charge. Volta also created the first transmission of electricity by linking positively charged and negatively charged. Connectors and driving an electrical charge or voltage through them.

In 1831 electricity become viable for use in technology when Michael Faraday created the electric dynamo [A crude Power generator] which solved the problem of generating electricity of electric current in an on-going and practical way. Faradays rather crude invention used a magnet that was moved inside a coil of copper wire creating a tiny electric current that flowed through the wire. This opened the door to American Scientist Thomas Edison and British scientist Joseph Swan who each invented the incandescent filament light bulb in their respective countries in about 1878 previously, light bulb had been invented by others, but the incandescent bulb was the first practical bulb that would light for hours on end. Swan and Edison later set up a joint company to produce the first practical filament lamp and Edison used his direct current system (DC) to provide power to illuminate the first New York electric street lamps in September 1882. Later in the 1800's and early 1900's Serbain American engineer, inventor and all around electrical wizard Nikola Tesla become an important contributes to birth the commercial electricity. He worked with Edison and later had many revolutionary developments in electromagnetism and had competing patents with Marconi for the invention of radio. He is well known for his work with alternating current (AC) AC motors and the poly phase distribution system.

Later American inventor and industrialist George westing house purchased and developed Teslas patentea motor for generating alternating current, and the work of westing house. Tesla and others gradually convinced American society that the future of electricity day with AC rather than DC. Others who worked to bring the use of electricity to where it is today include Scottish inventor James watt, Andre Ampere, A French mathematician and German mathematician and Physist George Ohm. And so it was not just one person who discovered electricity. While the concept of electricity was known for thousands of years when it came time to develop it commercially and scientifically there were several great minds working com the problem at the same time. For these whole genius scientists hard work our universe is able to run by electricity.

Electricity is the set of physical phenomena associated with the presence and motion of matters to that have a property of electric charge. Electricity is related to magnetism, both being part of the Phenomenon of electromagnetism as described by max wells equations. Various common Phenomena are related to electricity including lightning static electricity, electric heating, electric discharges and many other. The presence of an electric charge, which can be either positive or negative, produces an electric field. The movement of electric charges is an electric current and produce a magnetic field. Electricity is formed by Magnetic flux. The Question is how can magnetic flux form.

This magnetic flux is formed by a device which is name cald Transformer. We can say this is a heart of Electronics. This is similar to the human heart. The transformer is basically a voltage control device that is used widely in the distribution and transmission of alternating current.

Note- This is a very wide topic we can discoss about transformer in later topic.

Hear the question arise is how this transformer is similar to human heart.

We know that the heart is the main organ of human body of cardiovascular system, a network of blood vessels that pumps blood throughout body. This is also a very wide topic but the mechanism of transformer in electronics and the mechanism of human heart is too much similar we can discuss this topic in very easy form in later topic.

- Electric power where electric current is used to energies equipment.

- Electronics which deals with electrical circuits that involve active electrical components such as resisters, capacitors, Transistors, Diode and I.C Integrated circuits. This is very important components of any type of modern technologies. Because this is a direct responsible for co-operation, co-ordination, control, A complete management and Directed by I.C. is a brain of every component controlling in modern technologies.

Note- In this process the complete organization process is formed this is also a very important process in every modern technologies. We will discuss Later. And also I.C. is like a brain of we can say SOC System on the chip we will also discuss with human brain and microprocessor.

The theory of electromagnetism was developed in 19th century, and by the end of that century electricity was being put to industrial and residential use by electrical engineers.

Note- we will discuss later How is lower generating the rapid expansion in electrical technology at this time transformed industry and society, become a driving force for the second industrial revolution. Electricity extraordinary versatility monk it can be put to an almost limitless set of applications.

Chapter 9
Magnet

A Piece of iron, steel etc. that can attract and pick up other metal objects. Magnet is defined as a material that can produce its own magnetic field. There are three types of magnet. Permanent magnet, Temporary magnet and Electro magnet.

Magnets are the naturally occurring substances with the property of attracting iron. It is found that naturally occurring rocks have the property of attracting small particles of iron. Hence, they are called as natural magnets. They are Permanent magnets in nature. Iron, Nickel and cobalt are called magnet materials. A magnet is a rock or a piece of metal that can pull certain types of metal towards itself. The force of magnets called magnetism, is a basic force of nature, like electricity and gravity. Magnetism works over a distance. This means that a magnet does not have to be touching an object to pull it. A magnet is a material or artificial setup that can produce a magnetic field arown it. Due to the magnetic field, a magnet can attract ferromagnetic materials (eg. iron filings and attract or repel any other magnet. magnets suspended through a string, always point toward the north-south direction. A magnet always comes with a pair of magnetic Poles which cannot be separated. These are often referred to as "North Pole" and "South Pole". Like Poles repel on another whereas opposite Poles attract. Some materials naturally behave like magnet where as it is also possible to manufacture artificial magnets.

A magnet is an associate degree object or factor that produces a force field. This field of magnetism is invisible, however, it's liable for the magnetic field: the force that pulls like iron, steel, nickel, Cobalt and others. An eternal magnetic material that produces its own permanent force field.

Magnets can be categorized into the following

1- **Electromagnet-** Electromagnets are strong magnets, consisting of wires closely wrapped around an iron core. When a current is made to flow through the wires, it behaves like a magnet. As soon as the current is switched off, the magnetic behaviour goes away.

2- **Permanent Magnets-** Permanent magnets are products of solid magnetic force materials like alloy and primary solid solution, that are subjected to special process in an exceedingly sturdy force field throughout construction to align their crystal lime internal structure creating them very trouble. Some to get rid of magnets. A selected force field should be provided to get rid of the force field and this limit is set by the force of the force field. However due to high temperature and stress, even Permanent magnets can lose magnetic properties.

3- **Temporary Magnet**- These magnets are manufactured by exposing ferromagnetic materials to a magnetic field. When the magnetic field is removed, the materials lose their characteristics of a magnet. These magnets are made up of various soft materials

Properties of Magnet -

Magnets have distinctive and interesting Properties

1- **Attractive Property of Magnet-** A magnet attracts ferromagnetic materials like iron, nickel and cobalt.

2- **Directive Property of a magnet-** If a magnet is suspended from rigid support such that it can rotate freely, the magnet always points towards the north-south direction.

3- **Poles of a Magnet-** Magnets have two poles where the strength of the magnetic field is the strongest Magnetic poles exist in Pairs. No matter how small a magnet is this is a natural property of any magnet, it is impossible to separate one Pole. Like Poles always repel each other but opposite Poles attracts.

4- **The magnetic force-** Attraction or repulsion between two objects is inversely proportional to the distance between them. The force is stronger when the objects are close.

Applications of Magnet

1- **Speakers and microphones-** Several speakers use a static magnet and a coil that currently holds to convert power (signal) into mechanical. (noise-causing motion). The coil would hound the reel connected to the speaker come holds the signal as a versatile current that interacts with the static magnet field.

2- Like speakers, some electrical motors use a mixture of electrical magnets and permanent magnets to convert Power into energy. A generates convert energy into power by moving the conductor to a force field.

3- In addition to invasive surgery, hospitals use resonance imaging to diagnose the patient's limbs.

4- Nuclear resonance could be a methodology employed by chemists to provide products.

5- Magnets are widely used in daily life science and technology. Some uses are permanent magnets are used in hard drives, TV, cars, motors.

6- Temporary magnets are often used in manufacturing electromagnets.

We say magnets it a universally used in everywhere in many different-different ways in our daily life.

Did You Know?

The Earth has a magnetic field. It can be considered as a very large bar magnet with it's north pole being located near the geographical north pole and its south pole being located near the geographical south pole. Due to this magnet all magnets on the Earth Point in the geographic north-south direction.

The First magnetic material was discovered in the magnesia region of Asia. It was named Magnetic (Fe_3O_4). There is another story that suggests that the material was discovered by a shepherded named magnes who observed his shoe nails to stick with some rocks. Previously, magnetite was used by sailors in the oceans, as a compass. Therefore it is also known as leaging stone or lade stone.

Chapter 10
Basic Structure of Electric System

Power Generation

Electricity generation is the process of generating electric power from sources of primary energy. For utilities in the electric power industry, it is the stage prior to its delivery (transmission, distribution etc.) to end users or its storage (using for example the pumped storage method, Electricity is not freely available in nature, so it must be produced that is transforming other forms of energy to electricity. Production is carried out in power station also called power plants. Electricity is most often generated at a power plant by electromechanical generators, primary drived by heat engines fueled by combustion or nuclear fission but also by other means such as kinetic energy of flowing water and wind. Other energy sources include solar Photovoltaic and geothermal power. There are also exotic and speculative methods to recover energy, such as proposed fusion reactor designs which aim to directly extract energy from intense magnetic fields generated by fast moving charged particles generated by the fusion reaction.

Phasing out Coal- Field power stations and eventually gas fired power stations of capturing their green house gas emission is an important part of the energy transformation required to limit climate change. Vastly more solar power and wind power is forecast to be required with electricity demand increasing strongly with further electrification of transport homes and industries.

Michal Faraday discovered the fact that electric current can be produced by passing a magnet through copper wire. Today we are living in such an advanced world but almost all the electricity we are use today is produced with magnets and copper wires.

James Cleark, Max Well developed different equations that described electromagnetic fields in 1873. His prediction of the existence of electromagnetic waves traveling with the speed of light was conformed by Heinrich R. Hertz and in 1895, Marconi used these principles to make the use of electromagnetic waves to make Radio.

Development of alternating current A.C. was the turning point in the further journey of electricity. A Croatian scientist named Nikola Tesla worked with Thomas Edison. He came from the United States. Tesla discovered the Alternating current electrical systems with rotating magnets. Which is one of the most used principles today.

Electricity was a triller element for the scientists to work with, while led to the invention of so many devices and principles. Through all the devices are now more advanced and have more features, still the Principles are same and are being used. In large Power Plants to produce electricity. People wanted a cheap and safe way to light their homes, and scientists thought electricity might be a way.

A Different Kind of Power: The Battery Learning how to produce and use electricity was not easy. For a long time there was no dependable source of electricity for experiments. Finally, in 1800 Alessandro Volta, an Italian Scientist, made a great discovery. He Soaked Paper in salt water, placed zinc and copper on opposite sides of the paper and watched the chemical reaction Produce an electric current. Volta had created the first electric cell.

By connecting many of these cells together Volta was able to "string a current" and create a battery. It is in honor of Volta that we rate batteries in volts. Finally, a safe and dependable source of electricity was available, making it easy for scientists to study electricity.

A Current Began.

An English Scientist, Michael Faraday, was the First one to realize that an electric current could be produced by passing a magnet through a copper wire. It was an amazing discovery. Almost all the electricity we use today is made with magnets and coils of copper wire in giant power plants. Both the electric generator and electric motor are based on this Principle. A generator converts motion energy into electricity. A motor converts electrical energy into motion energy.

In 1879, Thomas Edison focused on inventing a practical light bulb, one that would last a long time before burning out. The problem was finding a strong material for the filament, the small wire inside the bulb that conducts electricity. Finally, Edison used ordinary cotton thread that had been soaked in carbon. This filament didn't burn at all-it became incandescent that is, it glowed. The Next challenge was developing an electrical system that could provide people with a practical source of energy to power these new lights. Edison wanted a way to make electricity both practical and inexpensive. He designed and built the first electric power plant that was able to produce electricity and carry it to people's homes. Edison's Pearl street Power Station started up its generator on September 4, 1882, in New York city. About 85 customers in lower Manhattan received enough lower to light 5,000 lamps. His customers paid a lot for their electricity, through. In today's dollars the electricity costs $5.00 Per Kilowatt-hour! Today, electricity costs about 13 cents per kilowatt-hour for residential customers, about 10.7 cents for commercial, and about 6.8 cents Per Kilowatt-hour for industry.

The Question : AC or DC?

The turning Point of the electricity age came a few years later with the development of AC [Alternating Current] power systems. Croatian born Scientist, Nikola Tesla came to the United States to work with Thomas Edison. After a falling out, Tesla discovered the rotating magnetic field and created the alternating current electrical system that is used very widely today. Tesla teamed up with engineer and business man George Westing house to patent the AC system and proved the nation with power that could travel long distances - a direct competition with Thomas Edison's DC system. Tesla later went on to form the Tesla Electric company, invent the tesla coil, which is still used in science labs and in radio technology today, and design the system used to generate electricity at Niagara Falls. Now using AC, Power Plants could transport electricity much farther than before. While Edison's DC (Direct current) Plant could only transport electricity within one square mile of his Pearl Street Power Station. The Niagara Falls Plant was able to transport electricity over 200 miles! Electricity didn't have an easy beginning. While many People were thrilled with all the new inventions, some people were afraid of electricity and wary of bringing it into their homes. They were afraid to let their children near this strange new power source. Many social critics of the day saw electricity as an end to a simpler, less hectic way of life. Poets commented that electric lights were less romantic than gaslights. Perhaps they were right, but the new electric age could not be dimmed. In 1920, about two percent of U.S energy was used to make electricity.

Q.1 What is electricity in the First Place?

Ans- Electricity simply refers to the movement of electrons through a conducting material, such as a copper wire. The force applied to electrons to push them through the conducting wire is known as voltage and the rate of flow of the electrons is known as current.

If you imagine the conducting wire as a pipe through which water can flow, voltage is the pressure applied to make the water flow while the current is how much water is flowing through the Pipe every second. In metals, electrons are free to move, which makes them great conductors of electricity. Some materials, however do not conduct electricity these are insulators. However there are instances when an insulator can carry an electrical charge. If you rub two different insulating materials together, such as balloon and Jumper, electrons will transfer from the jumper to the balloon. Which get loaded with a negative charge. This build-up of electrons on an insulators is known as static electricity- If you touch the balloon, you can feel there physics in action with a mild Shock.

Chapter 11
What is Electricity?

Electricity Keeps the Lights ON

Electricity is all around us. Whether it's our bedroom lamp, our favorite gaming system, or the fridge that holds all of our favorite snacks, electricity powers them all. These days it even powers many of our cars. You could travel to the most uninhabited areas of the Earth and still find it in the clouds. Above you during a storm. Electricity takes different forms: Coal, water, wind, solar, nuclear, and hydro. But have you ever wondered what exactly electricity is made of or how it manages to get to your house? Learning and understanding where electricity comes from and how we are able to consume it allows us to better manage our usage and be more mindful of our resources.

Q.1 What is electricity and how is it made?

Ans- Electricity is a secondary energy source that we get from the conversion of other sources of energy such as coal, natural gas, oil, nuclear power and so on. These sources are known as "Primary Sources". Primary sources can be renewable or non-renewable, but the electricity itself is neither. Like everything else, electricity is made up of atoms. So to understand electricity, it helps to understand basic information about atoms...

At the Center of an Atom is the nucleus. The nucleus is made up of particles called Protons and neutrons. Electrons revolve around the nucleus in shells. The Proton and electrons of an atom are attracted to each other and each carries an electrical charge. Protons have a positive charge and electrons have a negative charge. The positive charge of the Protons is equal to the negative charge of the electrons, making the atom balanced when they have an equal number of Protons and electrons. Neutron carry no electric charge and their number may vary.

Electrons

The electrons in the shell closest to the Nucleus have a strong attraction to the Protons. Sometimes the electrons in an atom's outermost shells do not have a strong attraction to the Protons and can be pushed out of the in orbits causing them to shift from one atom to another. These shifting electrons are electricity.

Q.2 How does electricity work?

Ans- Electricity travels in closed circuits. It has to have a complete path before elections can move through it. When you turn ON a light by flipping a switch, you close a circuit. Of course, this means that by flipping a switch off, you open a circuit. Electricity flows from the electric wire, through the light, and back into the wire. The same concept applies to your television or your appliances. When you turn them ON, you close a circuit for electricity to flow through the wires and Power them.

1- **Sources of electrical Energy** - Coal

More than one-fourth of the total known world coal reserves are in the United States. Through our dependence upon coal is decreasing, we still rely on it to produce electricity. Coal-generated electricity is created using a "pulverized coal combustion System" (PCC).

Coal is milled into a fine powder and is blown into a combustion chamber of a boiler and burned at a high temperature. The gases and heat energy produced converts water into steam. This steam passes through a turbine containing thousands of propeller like blades. At the end of these propellers, a generator sits mounted at one of the turbine shafts. When the generator's coils are rotated at a strong magnetic field, electricity is created.

The electricity generated is transported at higher Voltages via power lime grids. By the time it reaches our homes, the electricity is transformed down to safer 100 to 250 voltage systems.

2- Wind is a sources of electrical energy.

Wind energy is renewable and harnesses the energy generated by wind through the use of wind turbines that convert it into electricity. Wind, technically, is a by product of differences in temperature and is generated from the uneven heating of the atmosphere mountains, valleys and the revolution of the planets around the sun.

Wind turbines work the opposite way that fans do instead of using electricity to create wind, wind turbines use wind to make electricity. The wind turns the blades which spin a shaft that is connected to a generator and produce electricity.

3- Nuclear is a sources of electrical energy.

Nuclear energy comes from the energy in the core of an atom. Power plants use a Process called "nuclear Fission" - the splitting of an atom - to create energy. Some nuclear power plants use uranium atoms which are split when they are hit by a neutron, releasing heat and radiation creating more neutrons.

Once neutrons collide with other uranium atoms, the process repeats itself all over again. This chain reaction is controlled in nuclear power plants to produce heat. When combined with water, the heat produces steam which is then used to generate electricity that we can use in our homes.

4- Solar is a sources of electrical energy.

Solar energy uses the sun's light and heat to generate renewable or green energy. The most common forms of solar energy are harnessed by solar panels or photovoltaic cells. When rays hit the solar panels, it loosens electrons from their atoms and allows electrons to flow through the cell and generate electricity. In other ways, the Sun's energy is used to boil water and operate a steam turbine to generate electricity in a similar fashion to coal or nuclear Power Plants.

5- Hydro is a sources of electrical energy.

Hydroelectricity is created in a similar way that electricity from coal is created. In both cases, they require a power source to turn a turbine, which then turns a metal shaft in a generator that produces electricity. The biggest difference is where coal fired plants would use steam to turn their turbine blades, hydroelectricity plants use high pressure ducted water located at the base of dams to powes turbines.

Q.3 How does electricity get to your home?

Ans- Where electricity comes from and where consumers get their energy varies. Some utility companies get their energy varies generate all the energy they sell only using the lower plants they own. Others may purchase electricity directly from other utility companies, power marketers and independent power producers from a wholesale market organized by a regional transmission reliability organization. All the little details aside, electricity reaches the consumer in very similar ways.

Q.4 The electricity delivery process.

Ans- The electricity that Power Plants generate in delivered to customers over transmission and distribution power lines. The this complex system sometimes called the "grid" includes substations, Transformers, and power lines that connect electricity Producers and consumers. In the United States alone, the electricity grid contains thousands of miles of high- Voltage Power lines and millions of low-voltage Power lines with transformers that connect thousands of Power Plants to millions of consumers across the country. High Voltage Power lines that are hung between large metal towers are able to carry electricity over long distances, whereas lower voltage electricity is transmitted through transformers which increase or decrease voltages to adjust to different stages of the journey them the power plant to your home or business organization.

Q.5 What's a Smart meter?

Ans- The difference between one month's reading and the next is the number of energy units that have been used that billing Period. To help you save on your energy bill, smart meters are being installed all over the country. They provide two-way communication between you and your utility company, helping your utility known about events such as blackouts. It allows the utility to maintain more reliable services and can be used with home energy management systems such as web- based tools that your utility provides. They can even allow you to remotely adjust your thermostat or turn appliances off.

Q.6 Which Power Sources contributes to Pollution?

Ans- Although the cheapest form of generating fewer is through the burning of fossil fuels such as coal, natural gas, and oil, it is also the most hazardous to the environment. The burning of these fuels releases carbon dioxide into the atmosphere which has been linked to serious environmental complications including global warming, climate change, air pollution, natural disasters, habitate destruction, and chronic health problems.

Q.7 Which sources of powes are green?

Ans- Energy generated through renewable sources such as hydro, wind, solar and geothermal is green. Unlike fossil fuels, these sources of power do not deplete natural resources. They are also cleaner sources of energy that do not pollute the environment with carbon emissions.

Although renewable energy sources are better for the health of our planet, they typically cost more than other sources of energy, which is why the majority of our electricity is not generated from green sources.

Just Energy's Just Green Power Product makes it Possible for you to ensure that the equivalent of up to 100% of your electricity consumption is generated them renewable sources.

Although Just Energy's green energy options are available in most of the markets we serve, they are not yet available in all of our markets. See which markets we currently offer green energy options in.

Seven Unknown Facts about the Father of Electricity

1- Michal Faraday is responsible for the extensive study of electromagnetism, a field that changed the way human beings lived on this Planet. In order to pay homage to him "Farad" a unit used to measure electrical capacitance, is name after Faraday.

2- In 1824 Faraday invented the rubber balloon. He made balloon by cutting two sheets of rubber and pasting together the edges. He then filled the balloon with hydrogen. These balloons were used for his experiments on electro magnetism.

3- In 1826, Faraday had founded the Royal Institution of London's famous Friday Evening Discourses and the Christmas Lectures. Both of these Practices continue to this day.

4- Electromagnetic induction, the Principle behind the electric transformer and generator was discovered by Faraday in 1831.

5- It was Faraday who first popularized the concept of artificial cooling of refrigeration. He said that any gas can be pressurized to its liquid form and then released as vapour, which would make the gas act as a coolant.

6- Benzene, an important Petro-chemical product used in making plastic, was discovered by Michael Faraday. He found it in the oily residues of the gas lights in London.

7- Ever wondered why you don't get electrocuted while on a flight inside a storm? It's because the plane is built on the concept of Faraday cage. Michael Faraday invented an enclosure that could stop any electric charge from getting inside, protecting the object in the cage.

Chapter 12
Transformer is a Heart of Alternating Current

Electricity is all around us. But what is the mechanism of electricity which we can use easily in our daily life. In our early topics we understand the various types of electricity generating. In all these types at generating electricity a device is very importantly use for managing and transmitting the power from the power plant to our homes and working places. That device is called transformer. In the simplest way can be describe as a thing that step up or step down voltage. When we say or compare by human heart is almost very much similar functionality. We can say transformer is like a heart of Alternating current formation. And the Idea of transformer was first discussed by Michael Faraday in the year 1831 and was carried forward by many other prominent scientific scholars. However the generate purpose of using transformed was to maintain a balance between the electricity that was generated at very high voltages and consumption which was done at very low voltages. A Transformer is a device in the power transmission of electric the energy. As like human heart is the main organ of the human body of cardiovascular system. A network of blood vessels that Pump blood throughout the body. In same way the principal of transformer is the pumping of electricity when power is generated at very high voltage in power plant. The step-up transformer will decreases the output current and the step-down the transformer will increase the output current for keeping the input and the output power of the system equal. The transformer is basically a voltage control device that is used widely in the distribution and transmission of alternating current. In the same way the mechanism of human heart work with other body system to control heart rate and blood pressure. In other words we can say the heart is a first sized organ that pump blood throughout our body in control way. In the same way like this process is happens first in the alternating current is formed in power generating plants the power is not directly provided to the consumers firstly a step-up transformer is used for the output voltage is increase and in a step-down transformer the output voltage is decrease. For keeping the input and output power of the system equal. The transformer is basically a voltage control device that is used widely in the transmission and distribution of alternating current power. We can say it is a primary part of electricity, as like human heart which is the primary organ of humans circulatory System. Human heart consist of four main chambers (sections) made of muscles and powered by electrical impulses. Our brain and nervous system direct our heart's functions. The inside and outside of our heart contains components that direct blood flow. The main function of our heart is to move blood throughout our body. In the same way the main function of transforms is basically flowing of current in control way. That is used widely in the distribution and transmission of alternating current and however the general purpose of using transformer was to maintain a balance between the electricity that was generated at very high voltages and consumptions which was done at very law voltages. Basically a transformer is a device in the power transmission of electric energy. The transmission current is AC. It is commonly used to increase or decreases the supply voltage without a change in frequent of AC between circuits. In the same manner human heart is also controls the rhythm and speed of our heart rate. Maintains of blood pressure our heart work with other body systems to control hear rate and other body functions. The primary system of heart is. [The mechanism of heart working]

Nervous system- Our nervous system helps control but heart rate. It sends signals that tell our heart to beat slower during rest and faster during stress.

Endocrine System- Our endocrine System sends our harmons. These harmons tell our bleed vessels to constrict or relax which affects our blood pressure. Harmons from our thyroid gland can also tell our heart to beat faster or slower our heart is located in the front of our chest. It sits slightly behind and to the left of our sternoun (brest bone) our rib cage protects our heart. Our heart is slightly on the left side of our body. It sits between our right and left lungs. The left lung is slightly smaller to make room for the heart in our left chest. Every one heart is slightly different size. Generally adult heart are about the same size as two clenched fist. And children's heart are about the same size as two clenches fists and children's heart are about the same size as one clenched fist. On average an adult heart weight about 10 ounces our heart may weight a little more or a little less, depending on our body size and sex. The parts of our heart are like the parts of a house.

1) Walls, 2) Chambers (Rooms) 3) Valves (Doors) 4) Blood Vessels (Plumbing) 5) electrical conduction system (electricity)

Heart walls- Our heart walls are the muscles that contracts (squeeze) and relax to send blood throughout our body. A layer of muscular tissues called the septum divider our heart walls into the left and right sides. From the Transformer working principal is based on the electromagnetic induction and mutual induction. Transformers are used in various fields like power generation grid, distribution sector transmission and electric energy consumption. There are various types of transformers which are classified based on the following factors.

1- Working Voltage range
2- The medium used in the core
3- winding arrangement
4- Installation location.

Based on voltage Level

Commonly used transformer type, depending upon voltage they are classified as

1- **Step up-Transformer-** They are used between the power generator and the power grid. The secondary output voltage is higher than the input voltage.

2- **Step down Transformer-** These Transformers are used to convert higher voltage primary supply to low voltage secondary output. Based on the medium of core used in a transformer we will find different types of cases that are used

3- **Air-Core Transformer-** The flux linkage between primary and the secondary winding is through the air. The coil or windings wound on the non-magnetic strip.

4- **Iron core Transformer-** Windings are wound on multiple iron plates stacked together which provides a perfect linkage Path to generate flux.

Based on the winding Arrangement.

Auto Transformer- It will have only one winding wound over a laminated core. The Primary and secondary share the same coil. Auto also means self.

Based on Install Location

1- **Power Transformer-** It is used at power generation station as they are suitable for high voltage application.

2- **Distribution Transformer-** Mostly used at distribution lines for domestic purposes they are designed for, carrying low voltages. It is very easy to install and characterized by low magnetic losses.

3- **Measurement Transformer-** These are further classified. They are mainly used for measuring voltage, current, power.

4- **Protection Transformer-** They are used for components Protection Purposes. In circuits some components must be protected from voltage fluctuation etc. Protection transformers ensures components protection.

Working principal of transformer.

The Transformer works on the Principal of Faraday's low of electromagnetic induction and mutual induction.

There are usually two coils Primary and secondary coil on the Transformer core. The core laminations are joined in the form of strips. The two coils have high mutual inductions. When an Alternating current pass through the primary coil it creates a varying magnetic flux. As per Faraday's low of electromagnetic induction, this change in magnetic flux induces an (EME) Electro Motive Force, in the secondary coil, which is linked to the core having a Primary coil. This is mutual induction overall, a Transformer carries the below operation.

1- Transfer of electrical energy from circuit to another.

2- Transfer of electrical power through electromagnetic induction.

3- Electric power transfer without any change in the frequency.

4- Two circuits are linked with mutual induction.

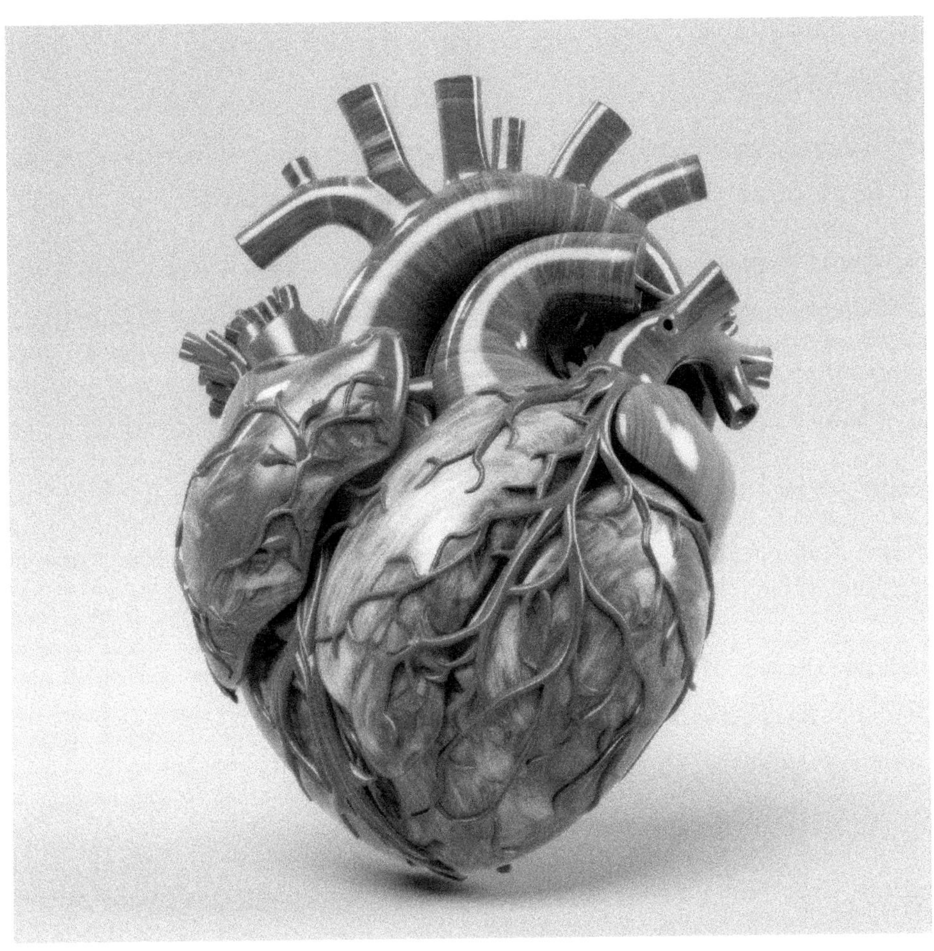

The figure shows the formation of varying magnetic flux lines around a wire-wound the interesting part is that reverse is also true, when a magnetic flux line fluctuates around a piece of wire a current will be induced in it. This was what Michael Faraday found in 1831, which is the functionental working principal of electric generators as well as transformers.

Parts of a single phase transformer.

1- **Core-** The core acts as a support to the winding in the transformer. It also provides a low reluctance path to the flow of magnetic flux. The winding is wound on the core as shown in the picture. It is made up of a laminated soft iron core in order to reduce the losses in a transformer.

There is three type of core

1) E-Type of core, 2) I-Type of core 3) Bobin

E-Shape core is fixed in both side of bobin and in between the empty space I-shape core is fixed. The factor Such as operating voltage, current power etc. Decide care composition. The core diameter is directly proportional to copper losses and inversely proportional to iron losses.

2- **Winding**- Windings are the set of copper wire wound over the transformer core. Copper wires are used due to

1- The high conductivity of copper minimizes the loss in a transformer because when the conductivity increases, resistance to current flow decreases.

2- The high ductility of copper is the property of metals that allows it to made into very thin wires. There are mainly two types of windings Primary Winding and secondary winding.

I- **Primary Winding**:- The set of turns of windings to which supply current is fed.

II- Secondary winding :- The set of turns of winding from which output is taken.

The Primary and secondary windings are insulated from each other using insulation coating agents.

3- **Insulation Agents**- Insulation is necessary for Transformer to separate windings from each other and to avoide short circuit. This facilitates mutual induction insulation agents have an influence on the durability and the stability of Transformer.

Following are used as an insulation medium in a transformer.

Insulating oil, Insulating Tape, Insulating Paper, wood-based lamination.

Ideal Transformer- The ideal transformer has no losses. There is no magnetic leakage flux, ohmic, resistance in its windings are no iron loss in the core.

Applications of Transformers.

1- The Transformer transmits electrical energy through wires over long distances.

2- Transformers with multiple secondary's are used in radio, TV receivers which require several different voltages.

3- Transformers are used as voltage regulators.

Now we will understand about Human heart walls which have three Layers.

1- Endocardium - Inner layer

2- Myocardium - Muscular middle layer

3- Epicardium - Protective outer layer

The epicardium is one layer of our pericardium. The Pericardum is a protective sac that cover our entire heart. It produces fluid to lubricate our heart and keep it from rubbing agent other organ

Heart chambers- Our heart is divided into four chambers. We have two chambers on the top catrium, plural atria and two on the bottom (ventricles) One on each side of the heart.

1. **Right atrium-** Two large veins deliver oxygen- poor blood to our right atrium. The superior vena cava carries blood from our upper body. The interior vena cava brings blood from the lower body then the right atrium pump the blood to our right ventricle.

2. **Right ventricle-** The lower right chamber pumps the oxygen-poor blood to our lungs through the pulmonary artery. The lungs reload blood with oxygen.

3. **Left atrium-** After the lungs fill blood with oxygen, the pulmonary veins carry the blood to the left atrium. This upper chamber pumps the blood to our left ventricle.

4. **Left Ventricle-** The left ventricle is slightly larger than the right. It pumps oxygen-rich blood to the rest of our body.

Heart valves- Our heart valves are like doors between our heart chambers they open and close to allow blood to flow through.

The atrioventricular valves (AV) open between our upper and lower heart chambers. They include

1- **Tricuspid valve-** Door between our right atrium and right ventricle.

2- **Mitral valve-** Door between our left atrium and left ventricle.

Semilunar valves (SL) open when blood flows out of our ventricles. They includes.

Aoritic valve: Open when blood flows out of our left ventricle to our aorta artery that carries oxygen-rich blood to our body.

Pulmonary valve: Open when blood flows from our right ventricle to our pulmonary arteries (The only arteries that carry oxygen-poor-blood to our lungs.

Blood vessels: Our heat pumps blood through three types of blood vessels.

Arteries- Carry oxygen-rich blood from our hear to our body's tissues. The exception is our pulmoner arteries, which go to our lungs, vein carry oxygen-poor blood back to our heart.

Capillaries- They are small blood vessels where our body exchanges oxygen rich and oxygen-poor blood.

Our heart receives nutrients through a network of coronary arteries. These arteries run along our heart's surface they serve the heart itself.

Left coronary artery: Divided into two branches. The circumflex artery and the left anterior descending artery.

Circumflex artery: Supplies blood to the left atrium and the side and back of the left ventricle.

1- Left anterior descending artery (LAD)- Supplies blood to the front and buttom of the left ventricle and front of the septum.

2- Right coronary artery (RCA) – Supplies blood to the right atrium right ventricle, bottom portion of the left ventricle, bottom portion of the left ventricle and lack of the septum.

Electrical conduction system- Our heart conduction system is like the electrical wiring of a house. It controls the rhythm and pace of our heart beat. It includes.

1- Sinoatrical node (SA) Node- Sends the signals that make our heart beat.

2- Atrioventricular node (AV) Node- Carries exlectrical signals from our heart upper chamber to its lower ones.

Our heart also have a network of electrical bundles and fibers. This network includes.

1- Left bundle branch: Sends electric impulses to our left ventricle.

2- Right bundle branch: Sends electric impulses to our right ventricle

3- Bundle of his : Sends impulses from our AV node to the Purkinje fibers.

4- Purkinje Fibers: Make our heart ventricles contract and pump our blood.

Sinoatrial Node (S.A node) is generate electric current to pump the heart. In one minute it beet 72 times. And at one time heart is beats it will Pumps70 ml of blood. In one time beating and in our body have 5 Five leater of blood. Which is pumped by our heart in one minutes. When we calculate the heart pumping time of 5000 ml blood is 72 times of heart beat × 70 ml of blood it comes 5,040 approx. And the capacity of heart is 50 ml. The blood pressure is the rate of pumping heart is 120 and Receiving is 80 the sound of heart is called heart beat. When the (S.A node) sinoatrial node is working. A pacemaker machine is fixed in right atrion by doing by pass sergury. This is a small device that's placed (implanted) in the chest to help control the heart beat. It's used to Prevent the heart from beating too slowly. A pacemaker is also called a cardiac pacing device.

As like our heart have 3 Three layers and 4 Four chambers. In comparison of heart and transformer we observes there is two windings the Primary windings and secondary winding.

Chapter 13
Transformer Basics

Transformers are electrical devices consisting of two or more coils of wire used to transfer electrical energy by means of a changing magnetic field. We will see that a transformer has no internal moving parts, and are typically used because a change in voltage is required to transfer energy from one circuit to another by electromagnetic induction. One of the main reasons that we use alternating currents. AC voltages in our houses and workplace's is that AC supplies can be easily generated at a convenient voltage, transformed (hence the name transformer) into much higher voltages and then distributed around the country using a national grid of pylons and cables over very long distances. The reason for transforming the voltage to a much higher levels is that higher distribution voltages implies lower currents for the same power and therefore lower I^2*R losses along the networked grid of cables. These higher AC transmission voltages and currents can then be reduced to a much lower, safer and usable voltage level where it can be used to supple electrical equipment in ur homes and workplaces, and all this is possible thanks to the transformer founder Michal faraday basics of the voltage Transformer

The voltage Transformer can be thought of as an electrical component rather than an electronic component. A transformer basically is very simple static (or stationary) electro-magnotic passive electrical device that works on the principle of faraday's law of induction converting electrical energy from one value to another.

The transformer does this by linking together two or more electrical circuits using a common oscillating magnetic circuit which is produced by the transformer itself. A transformer basic operate on the principles of "electromagnetic induction", in the form of mutual induction.

Mutual induction is the process by which a coil of wire magnetically induces a voltage into another coil located in close proximity to it. Then we can say that transformer work in the "magnetic domain", and transformer get their name from the fact that they "transform" one voltage or current level into another, Transformer are capable of either increasing or decreasing the voltage and current levels of their supply, without modifying its frequency, or the amount of electrical power being transferred from one winding to another via the magnetic circuit.

A single Phase voltage transformer basically consists of two electrical coils of wire, one called the "primary winding and the another called the "secondary winding" The "primary" side of the transformer as the side that usually takes power, and the "Secondary" as the side that usually delivers power. In a single-phase voltage transformer the primary is usually the side with the higher voltage.

These two coils are not in electrical contact with each other but are instead wrapped together around a common closed magnetic iron circuit called the "core" This soft iron core is not solid but made up of individual laminations connected together to help reduce the core's magnetic losses.

The primary and secondary windings are electrically isolated from each other but are magnetically linked through the common core allowing electrical power to be transferred from one coil to other. When an electric current passed through the primary winding, a magnetic field is developed which induces a voltage into the secondary winding.

Single phase voltage Transformer

In other words, for a transformer there is no direct electrical connection between the two coil windings, there by giving it the name also of an isolation transformer. Generally, the primary winding of a transformer is connected to the input voltage supply and convers or transforms the electrical power into a magnetic field While the Job of the secondary winding is to convert this alternating magnetic field into electrical power producing the required output voltage.

Transformer Construction (Single-Phase)

Vp- is the primary voltage

Vs- is the secondary voltage

Np- is the number of primary windings

Ns- is the number of secondary windings

Ø (Phi)- is the flux linkage

Notice that the tow coil windings are not electrically connected but are only linked magnetically. A Single phase transformer can operate to either increase or decrease the voltage applied to the primary winding. When a transformer is used to "increase" the voltage on its secondary winding with respect to the primary it is called a step-up-Transformer. When it is used to "decrease" the voltage on the Secondary winding with respect to the primary it is called a step-down transformer.

However, a third condition exists in which a transformer produces the some voltage on its secondary as is applied to its primary winding. In other words, its output is identical with respect to voltage, current and power transferred. This type of transformer is called an "Impedance transformer" and is mainly used form impedance matching or the isolation of adjoining electrical circuits. The difference in voltage between the primary and the secondary windings is achieved by changing the number of coil turns in the Primary winding (Np) compared to the number of coil turns on the secondary winding (NS) as the transformer is basically a linear device, a ratio now exists between the number of turns of the primary coil divided by the number of turns of the secondary coil. This ratio, called the ratio of transformation, more commonly known as a transformers "turns ratio" (TR) This turns ratio value dictates the operation of the transformer and the corresponding voltage available on the secondary winding.

It is necessary to know the ratio of the number of turns of wire on the primary winding compared to the secondary winding. The turns ratio, which has no units, compares the tow windings in order and is written with a colon, such as 3:1 (3- to – 1)

This means in this example, that if there are 3 volts on the primary winding there will be 1volt on the secondary winding, 3 volts-to-1volt. Then we can see that if the ratio between the number of turns changes the resulting voltages must also change by the same ratio, and this is true.

Transformers are all about "ratios" the ratio of the primary to the secondary, the ratio of the input to the output, and the turns ratio of any given transformer will be the same or its voltage ratio. In other words for a transformer. "turns ratio- voltage ratio" The actual number of turns of wire on any winding is generally not important, Just the turns ratio and this relationship is given as.

A Transformers Turns Ratio

$$\frac{N_P}{N_S} = \frac{V_P}{V_S} = n = \text{Turns Ratio}$$

Assuming an ideal transformer and the phase angles Ø_P -Ø_S

Note that the orders of the numbers when expressing a transformers turns ratio value is very important as the turns ratio 3:1 expresses a very different transformer relationship and output voltage than one in which the turns ratio is given as 1:3

Transformer Basics Example No. 1

A voltage transformer has 1500 turns of wire on its primary coil and 500 turns of wire for its secondary coil. What will be the turns ratio (TR) of the transformer.

$$\text{T.R.} = \frac{N_P}{N_S} = \frac{\#Pri.Coils}{\#\,Sec.Coils} = \frac{1500}{500} = \frac{3}{1} = 3{:}1$$

This ratio of 3:1 (3 to 1) simply means that these are three primary windings for every one secondary winding, As the ratio moves from a larger number on the left to a smaller number on the right, the primary voltage is therefore stepped down in value as shown.

Transformer Basic Example no.2

If 240 volts rms is applied to the primary winding of the same transformer above, what will be the resulting secondary no load voltage

$$\text{T.R.} = 3{:}1 \text{ or } \frac{3}{1} = \frac{V_P}{V_S} = \frac{\#Pri\ Volts}{\#\ Sec\ Volts} = \frac{240}{V_S}$$

$$\therefore Sec.Volts\ \frac{V_P}{3} = \frac{240}{3} = 80 \text{ volts}$$

Again confirming that the transformers is a stepdown "Transformer as the primary voltage is 240 volts and the corresponding secondary voltage is lower at 80 volts.

Then the main purpose of a transformer is to transform voltages at present ratios and we can see that the primary winding has a set amount or number of windings (coils of wire) on it to suit the input voltage If the secondary output voltage is to be the same value as the input voltage on the primary winding, then the same number of coil turns must be wound on the secondary core as there are on the primary core giving an even turns ratio of 1:1 (1-to-1). In other words one coil turn on the secondary to one coil turn on the primary.

It the output secondary voltage is to be greater or higher than the input voltage, (step-up-transformer) then there must be more turns on the secondary giving a turns ratio of 1:N (1-to-N), where N-represents the turns ratio number. Likewise, if it is required that secondary voltage is to be lower or less than the primary (step-down transformer) then the number of secondary windings must be less giving a turns ratio of N:1 (N-to-1)

Transformer Action

We have seen that the number of coil turns on the secondary winding compared to the primary winding, the turns ratio, affects the amount of voltage available from the secondary coil, But if the two windings are electrically isolated from each other, how is this secondary voltage produced?

We have said previously that a transformer basically consists of two coils wound around a common soft iron core. When an alternating voltage (Vp) is applied to the Primary coil, current flows through the coil which in turns sets up a magnetic field around itself. This effect is called mutual inductance according to faraday's Law of electromagnetic induction.

The strength of the magnetic field builds up as the cassez flow rises from zero to its maximum value which is given as $d\phi/dt$.

The magnetic lines of force set up by this electromagnetic expand outward from the coil the soft iron core forms a path for and concentrates the magnetic flux. This magnetic flux links the turns of both windings as it increases and decreases in opposite directions under the influence of the AC supply.

However, the strength of the magnetic field induced into the soft iron core depends upon the amount of current and the number of turns in the winding. When current is reduced, the magnetic field strength reduces.

When the magnetic lines of flux flow around the core, they pass through the turns of the secondary winding causing a voltage to be induced into the secondary coil. The amount of voltage induced will be determined by: N*dϕ/dt (Faraday's Law), where N is the number of coil turns. Also this induced voltage has the same frequency as the primary winding voltage.

Then we can see that the same voltage is induced in each coil turn of both windings because the same magnetic flux links the turns of both the windings together. As a result, the total induced voltage in each winding is directly proportional to the number of turns in that winding However, the peak amplitude of the output voltage available on the secondary winding will be reduced if the magnetic losses of the core are high.

If we want the primary coil to produce a stronger magnetic field to overcome the cores magnetic losses, we can either send a larger current through the coil, or keep the same current flowing, and instead increase the number of coil turns (Np) of the winding. The product of amperes times turns is called the "ampere-turns", which determines the magnetizing force of the coil.

So assuming we have a transformer with a single turn in the primary, and only one turn in the secondary. If one volt is applied to the one turn of the primary coil, assuming no losses, enough current must flow and enough magnetic flux generated to induce one volt in the single turn of the secondary. That is, each winding supports the same number of volts per turn.

As the magnetic flux varies sinusoidally, ϕ-ϕ max Sinwt, then the basic relationship between induced emf, (E) in a coil winding of N turns is given by:

emf = turns × rate of change

$$E = N\frac{d\phi}{dt}$$

$$E = N\omega \times \phi_{max} \times \cos(\omega t)$$

$$E_{max} = N\omega\, \phi_{max}$$

$$E_{rms} = \frac{N\omega}{\sqrt{2}} \times \phi_{max} = \frac{2\pi}{\sqrt{2}} \times f \times nx\phi_{max}$$

$$\therefore E_{rms} = 4.44\, fN\phi_{max}$$

Where

f- is the flux frequency in Hertz

= ω/2π

N- is the number of coil windings

ϕ – is the amount of flux in webers.

This is known as the Transformer EMF Equation.

For the primary winding emf, N will be the number of primary turns, (Np) and for the secondary winding emf, N will be the number of secondary turns, (Ns) also please note that as transformers require an alternating magnetic flux to operate correctly, Transformers cannot therefore be used to transform or supply DC voltages or currents, since the magnetic field must be changing to induced a voltage in the secondary Winding. In other words, transformers. Do not operate on steady State DC voltage, only alternating or pulsating voltages.

If a transformer Primary winding was connected to a DC supply, the inductive reactance of the winding would be zero as DC has no frequency; so the effective impedence of the winding will therefore be very low and equal only to the resistance of the copper used. Thus the winding will draw a very high current from the DC supply causing it to over heat and eventually burnout, because as we know I=V/R

Transformer Basic Example No3

A single phase transformer has 480 turns on the primary winding and 90 turns on the secondary winding. The maximum value of the magnetic flux density is 1:1 T Where 2200 volts, 50 Hz is applied to the transformer primary winding. Calculating:

a- The maximum flux in the core.

$$E_{rms} = \frac{N\omega}{\sqrt{2}}\phi_{max}$$

$$\phi_{max} = \frac{E_{rms}}{N\omega} \times \sqrt{2} = \frac{2200}{480 \times 2\pi \times 50} \times \sqrt{2}$$

$$\therefore \phi_{max} = 0.0206 \text{ Wb or } 20.6 \text{ m Wb}$$

b- The cross- sectional areas of the core.

$$\phi_{max} = \beta \times A$$

$$\therefore A = \frac{\phi_{max}}{\beta} = \frac{0.0206}{1.1} = 0.0187m^2$$

c- The secondary induced emf

Es (rms) = 4.44 fNs ϕ_{max}

Es (rms) = 4.44 × 50 × 90 × 20.6 ×10^{-3}

∴ Es (rms) = 412 volts.

Since the secondary voltage rating is equal to the secondary voltage rating is equal to the secondary induced emf, another easier way to calculate the secondary voltage from the turns ratio is given as.

$$\text{T.R.} = \frac{V_P}{V_S} = \frac{N_P}{N_S}$$

$$\therefore V_S = \frac{V_P \times N_S}{N_P} = \frac{2200 \times 90}{480} = 412\ Volts$$

Electrical Power in a Trasnformer

Another one of the transformer basic parameters is its power rating. The power rating of the transformer is obtained a rating in volt- amperes (VA) small single phase transformers may be rated in volutampere only, but much larger power transformers are rated in units of kilo volt-amperes, (KVA) where I kilo volt-ampere is equal to 1,000 volt- amperes, and units of mega volt-amperes (MVA) where 1 mega volt-ampere is equal to 1 million volt-amperes.

In an ideal transformer (ignoring any losses), the power available in the secondary winding will be the same as the power in the primary winding, they are constant wattage devices and do not change the power only the voltage to current ratio. Thus, in an ideal transformer the power ratio is equal to one (unity) as the voltage, V multiplied by the current, I will remain constant. That is the electric power at one voltage/current level on the secondary side. Althrough the transformer can step-up (or step-down) it cannot step-up power. Thus, when a transformer steps-up a voltage, it step-down The current and vice-versa, so that the output Power is always at the same value as the input power. Then we can say that primary power equals secondary power, (Pp-Ps)

Power in a Transformer

$$Power_{Primary} = Power_{Secondary}$$

$$P_{(PRI)} = P_{(SEC)} = V_P I_P \cos\phi_P = V_S I_S \cos\phi_2$$

Where: ϕ_P is the primary Phase angle and ϕ_s is the secondary phase angle.

Note that since Power loss is proportional to the square of the current being transmitted, that is: I2 R, increasing the voltage, let's say doubling (×2) the voltage would decrease the current by the same amount (÷2) while delivering the same amount of power to the load and therefore reducing losses by factor of 4. If the voltage was increased by a factor of 10, the current would decrease by the same factor reducing overall losses by factor of 100.

Transformer Basics-Efficiency

A transformer does not require any moving parts to transfer energy. This means that there are no friction or windage losses associated with other electrical machines. However, transformers do suffer from other types of losses called "copper losses" and iron losses but generally there are quite small.

Copper losses, also known as I^2R loss is the electrical Power which is lost n heat as a result of circulating the current around the transformers copper windings, hence the name. Copper losses represents the greatest loss in the operation of transformer.

The actual watts of power lost can be determined (in each winding) by squaring the amperes and multiplying by the resistance in ohms of the winding (I^2R)

Iron losses, also known as hysteresis is the lagging of the magnetic molecules within the core, in response to the alternating magnetic flux. This lagging (or out-of-phase) condition is due to the fact that it requires power to reverse magnetic molecules; they do not reverse until the flux has attained sufficient force to reverse them.

Their several results in friction, and friction produces heat in the core which is a form of power loss. Hysteresis within the transformer can be reduced by making the core from special steel alloys.

The intensity of power loss in a transformer determines its efficiency. The efficiency of a transformer is reflected in power (wattage) loss between the primary (input) and secondary (output) windings. Then the resulting efficiency of a transformer is equal to the ratio of the power out put of the secondary winding, P_S to the Power input of the primary winding, P_P and is therefore high.

An ideal transformer would be 100% efficient, passing all the electrical energy it receives on its primary side to its secondary side. But real transformers on the other hand are not 100% efficient. When operating at full load capacity their maximum efficiency is nearer 94% to 96% which is still quite good for an electrical device. For a transformer operating at a constant AC voltage and frequency its efficiency can be as high as 98%. The efficiency, n of a transformer is given as:

Transformer Efficiency

$$\text{efficiency}, \eta = \frac{Output\ Power}{Input\ Power} \times 100\%$$

$$= \frac{Input\ Power - Losses}{Input\ Power} \times 100\%$$

$$= 1 - \frac{Losses}{Input\ Power} \times 100\%$$

Where: Input, output and losses are all expressed in units of power.

Generally when dealing with transformers, the primary watts are called "volt-amps", VA to differentiate then from the secondary watts. Then the efficiency equation above can be modified to:

$$\text{Efficiency, } \eta = \frac{Secondary\ Watts\ (Output)}{Primary\ VA\ (Input)}$$

When learning about transformer basics, it is sometimes easier to remember the relationship between the transformers input, output and efficiency by using pictures. Here the three quantities of VA, W and n have been superimposed into a triangle giving power in watts at the top with volt amps and efficiency at the bottom. This arrangement represents the actual position of each quantity in the efficiency formulas.

Transformer Efficiency Triangle and the transposing the above triangle quantities gives us the following combinations of the same equations:

Then, to find watts (output) = VA× eff., or to find VA (input)

= W/eff., or to find efficiency, eff. = W/VA, etc.

Notice- A like humans heart beet has some sound similarly when transformer is working or generating electricity that time transformer generate humming sound which is similar to human heart beat. This is caused by a phenomenon called magnetostriction. In very simple terms this means that if a piece of magnetic sheet steel is magnetized it will extend itself. When the magnetization is taken away, it goes back to its original condition as we know that A transformer has no moving parts, So why does it make humming sound?

There are actually multiple causes of transformer noise. The main one is the magnetostriction Effect. This is when the current that flows through the transformer's coils creates a magnetic field. The magnetic field then changes the dimensions of the transformer's iron core. The core expands and contracts with the alternating current, Which causes a humming sound. We can say that this is same as humans heart beat when muscles is expand and contracts.

As the transformer ages. The layers within the core of the transformer begin to break a part and spate from one another. This cause the vibrations to get louder.

We can transmit power further and more efficiently at high voltages.

Chapter 14
Electricity is Universal

Michael faraday discovered the fact that electric current can be produced by passing a magnet though copper wire. Today we are liging in such as advanced world but almost all the electricity we use today is produced with magnets and copper wires.

James cleark Max well developed different equations that described electromagnetic fields in 1873. His prediction of the existence of electromagnetic waves traveling with the speed of light was conformed by Heninrich R. Hertz and in 1895, Marconic used these Principles to make the use of electromagnetic waves to make radio Development of alternating current or Ac current was turning point in the further Journey of electricity. A creation scientist name Nikola Tesla worked with Thomas Edision. He came from the united Stafes. Tesla discovered the A.C electrical systems with roating magnets which is one of the most used principles Today electricity was a triler element for the scientist to work with which led to the invention of so many devices and principle Though all the devices are now more advanced and have more features, Still the principles are same and are being used in large power plants to Produce electricity.

Q.1 What is electricity in First Place?

Ans- Electricity simply refers to the movement of electrons through a conducting material such as a copper wire. The force applied to electrons to push them though the conducting wire is known as voltage and the rate of flow of the electrons is known as current.

If you imagine the conducting wire as a pipe through which water can flow, voltage is the pressure applied to make the water flow while current is how much water is flowing through the pipe every second.

In metals, electrons are free to move, which makes them great conductors of electricity. Some materials, however do not conduct electricity these are insulators. However there are instances when an insulators can carry an electrical charge. If you rub two different insulating materials together, such as balloon and jumpers, electrons will transfer from the jumper to the balloon which get loaded with a negative charge. This build-up of electrons on an insulator is known as static electricity- if you touch the balloon, you can feel these physics in action with a mild shock.

History of Battery

Q.2 Electricity in the ancient world: The story of the Baghdad "Battery"

Ans- To the best of our knowledge, the Greeks were the first to discover the nation of electrical charge over 2,600 years ago. They observed that rubbing fossizlized three resin,or amber, with animal fur caused it to attract dried grass. Essentially the Greeks had come across static electricity. We also know from ancient texts that Egyptians know that some species of electric fish could trigger shocks in the body. In fact the ancient Egyptions likely used the electric nile cat fish to treat headaches and nerve pain a practice known as inchthyoectroanalgesia that remained in medical use until the late 1600s. But without a doubt, the most amazing example of electricity an antiquity is the Baghdad battery. This pecular instrument was discovered by an expedition led by Dr. Wilhelm Koenig of the Iraq Museum in Baghdad in 1936. The finding consisted of a vase made of clay about 14 centimeters hight and with the largest diameter at 8 centimeters.

Dating suggests tha the artifact is about 2000 year old, from the Ist century AD, during a time when the region was occupied by the Parthian empire.

Although its appearance didn't seem out of the oridinary scientists quickly learned that there was much more the small clay pot once they peeked inside.

The vase contains a hollow clynder made of a sheet of copper of high purity. The lower end of the cylinder was covered with a piece of sheet copper while the inner bottom of the cylinder was covered with a layer of asphalt, only 3 millimeters thick. The upper end of the cylinder was plugged by a heavy and thick layer of asphalt. The center of the plug featured a solid piece of iron

At the time of discovery, koening recognized that the Jar and its odd metal structure where in a configuration that suggest it could have functioned as a wet cell battery. In fact it seems to have served no other purpose than that of generating a weak electric current.

Experiments conducted with replies of the Jar employing various acids found that a mixture of acetic acid (distilled vinegar) and grape fruit juice generated 0.5 volts for several days more such artifacts have been discovered along the year around sites in mordon day Iraq which were made by Parthians and sassanids however what purpose could these ancient batteries have served considering that no motors, light or any similar electric device have been found. One possible application of the Baghdad battery is for medical therapy as the Greeks and Romans of the time routinely employed the common electric ray to deliver electric shocks to patients for treating pain.

This lack of any apparent use for electrical current has led some to question whether these ancient Jars were actually used as battery. Instead, These could have been to store important documents in order to keep the moisture from damaging the papyrus.

What's more, since there is no record that Parthians nor any more anyone in the ancient world for that matter, possessed a formal theory of electricity the discovery of the batteries was likely an accident.

Chapter 15
Dry Cell

Q.1 What is Dry cell

Ans- A dry cell is one type of electric battery, which is generally used for the home and portable electronic devices. A battery is a device that consists of one or more electrochemical cell, which converts chemical energy into electrical energy. A dry cell is one of the electrochemical cells, developed by the German scientists "Carl Gassner" in 1886, after the development of wet zinc- carbon batteries by Georges leclanche in 1866. Modern dry cells were developed by yai sakizo, who is from Japan in the year of 1887. Nowadays, the most commonly used battery are dry cells batteries which vary from large flashlight batteries to minimized flashlight batteries and are mostly used in wrist watches or calculators. A dry cell is an electrochemical cell consisting of low moisture immobilized electrolytes in the form of a paste, which restricts it from flowing. Due to this, it is easily transportable working Principle and types of dry cells. Depending on the nature of the dry cell, it can be classified as a primary cell and the secondary cell. A primary cell is the one which is neither reusable nor rechargeable. Once the electrochemical reactions consume all the chemical reagents. They fail to produce electricity. On the other hand, a secondary cell can be rechargeable the chemical reactions.

Primary cell

1- **Zinc-Carbon Cell-** A dry cell consists of a metal container in which a low moisture electrolyte paste convers the graphite rod or a metal electrode. Generally, the metal container will be zinc whose base acts as a positive electrode (cathode). It is surrounded by managanese dioxide and low moisture electrolyte like ammonium chloride paste, which will produce a maximum of 1.5 v of voltage, and they are not reversible.

The half cell reaction process has the following steps.

Step.1- During the process, a reduction reaction occurs within the moisturized electrolyte, which comprises manganese dioxide (MnO_2) and ammonium chloride (NH_4cl) and graphite serves as solid support for the reduction reaction.

$2NH_4^+ + 2Mn_2O_3 + 2NH_3 + H_2O$

Step-2- Zinc container serves as an anode and undergoes an oxidation reaction.

$$Zn \rightarrow Zn^{2+} + 2^{e-}$$

Zinc- carbon cell is the most common dry cell and is also called leclanche cell. The alkaline battery has almost same half-cell reaction, Where KOH or NaoH replaces the ammonium chloride and half – cell reactions are

$ZNCl_2 + 2NH_3 \rightarrow Zn(NH_3)_2 Cl_2$

$2MnO_2 + H_2 - Mn_2O_3 + H_2O$

The overall reaction is

$Zn + 2MnO_2 + 2NH_4Cl \rightarrow Mn_2O_3 + Zn(NH_3)_2 Cl_2 + H_2O$

2- Alkaline battery

The alkaline battery will have almost same half-cell reactions as zinc-carbon cell, Where KOH or NaoH replaces the ammonium chloride and half-cell reactions are

$Zn + 2OH^- \rightarrow ZnO + H_2O + 2e^-$

$2MnO_2 + 2e^- + H_2O \rightarrow Mn_2O_3 + 2OH^-$

3- Mercury Cell

In the mercury cell, HgO serves as a cathode and zinc metal serves as an anode and the reaction involves the following steps

Step-1 at the anode

$Zn + 2OH^- \rightarrow ZnO + H_20 + 2e^-$

Step-2 at the cathode

$HgO + H_2O + 2e^- \rightarrow Hg + 2OH^-$

The over reactions of the cell

$Zn + HgO \rightarrow ZnO + Hg$

4- Silver oxide cell

In the basic medium, silver metal acts as insert support in the reduction of silver oxide (Ag_2O) and in the oxidation of zinc

Step-1 Reaction at the cathode

$Ag_2O + 2H^+ + 2e^- \rightarrow 2Ag + H_2O$

Step-2 Reaction in the electrolyte

$2H_2O \rightarrow 2H^+ + 2OH^-$

Step-3 Reaction at the anode

$Zn + 2OH^- \rightarrow Zn(OH_2) + 2e^-$

Step- 4 overall reaction

$Zn + H_2O + Ag_2O \rightarrow Zn\ (OH_2) + 2Ag$

The overall reaction in an anhydrous medium

$$Zn + Ag_2O \xrightarrow[NaOH]{KOH} ZnO + 2Ag$$

Secondary Cell

Nickel- Cadmium Cell (Nicd cell)

The Nickel- Cadmium Cell compises cadmium as an anode and a nickel plate as a cathode and a separator acts as an insulator between the anode plate and cathode plate. Sodium hydroxide or potassium hydroxide acts as an electrolyte.

Note: On disposal, cadmium causes harmful effects on the environment. Therefore, these days Nicd cells are not in use

2- Lithium-ion cell- These are popular batteries used now a days on laptops, ipads and cellphones. The electrodes of the cell are made up of light weight carbon and lithium. They are low-maintanance batteries and no memory is required to enhance battery life. They are less harmful even after disposing and self discharge is less than half of the Nicd cell

3- Nickel-Metal Hydride cell

In the nickel-metal hydride cell, NIMH acts as an anode and hydrogen absorbing alloy acts as a cathode. The electrochemistry of the rechargeable Ni-MH battery is a following.

Step-1: Reaction at the chathode

$H_2O + M^+e^- \leftrightharpoons OH^- + MH$

Step-2: Reaction at the anode

$Ni(OH)_2 + OH^- \leftrightharpoons NIO(OH) + H_2O + e^-$

And at the end of the reaction, nickel oxhydroxid NiO(OH) is formed

Advantages of a Dry cell

1- They are easy to use in simple electronic device

2- There is no worry about linkage in the dry cell

3- Dry cell are safe during transportation

4- They are small and lightweight.

Cell structure unit of Life

Q.1 What is cell?

Ans- It is defined as the cell is the smallest functional and structural unit of life then the Group of cells form tissues and the group of tissues of same function form organ. That organ form system by the system form full body. Robert book first I[st] discovered the cell in cork.

Anton Von Leuwenhoek first I[st] observed live cell.

Structure of cell

In the structure of cell there are following organelles:-

Cell membrane, cytoplasm, Golgi apparatus, Lysosome, Mitochondria, Ribosomes, Nucleus.

Cell membrane- It is outer protecting covering of cell, cell membrane is mainly consist of three component

(i) Protein 55% (ii) Lipid 40% (iii) Carbohydrates 5%

Cytoplasm- The fluid present inside the cell, formed by 80% of water (80 Percent). In cytoplasm cell organelles float.

Nucleus- It is very important cell organ.

It is a largest cell organelle.

It is also called controller of cell.

It consist of nuclear membrane.

It is first I[st] discovered by Robert Brown.

It is coordinator.

Cell

Cells are the basic, fundamental unit of life. So if we were to break apart an organism to the cellular level, the smallest independent component that we would find would be the cell. Explore the cell notes to know what is the Cell, cell defination, cell structure, types and functions of cells. These notes have an indepth description of all the concepts related to the cells.

Q.1 What is a cell?

Ans- A cell is defined as the smallest basic unit of life that is responsible for all of life's processes.

Cell are the structural, functional, and biological units of all living beings. A cell can replicate itself independently. Hence, they are known as the building blocks of thy life. Each cell contains a fluid called the cytoplasm. Which is enclosed by a membrane also present in the cytoplasm are several biomolecules like Proteins, nucleic acids and lipids. Moreover, cellular structures called cell organelles are suspended in the cytoplasm.

A cell is the structural and fundamental unit of life. The study of cells from its basic structure to the functions of every cell organelle is called cell Biology. Robert Hooke was the first I[st] Biologist who discovered cells. All organisms are made up of cells. They may be made up of a single cell (Unicellular) or many cells (multicellular). Mycoplasmas are the smallest known cells. Cells are the building the blocks of all living beings. They Provide structure to the body and convert the nutrients taken from the food into energy.

Cells are complex and their components perform various functions in an organism. They are of different size and shapes. Pretty much like bricks of the buildings. Our body is made up of cells of different shapes and size.

Cells are the lowest level of organization in every life form. From organism to organism, the count of cells may vary. Humans have more number of cells compared to that of bacteria. Cells comprise several cell organelles that Perform specialised functions to carry out life processes. Every organelle has a specific structure. The hereditary material of the organisms is also present in the cells.

Discovery of cells

Discovery of cells is one of the remarkable advancements in the field of science. It helps us know that all the organisms is made up of cell and these cells help in carrying out various life processes. The structure and function of calls helped us to understand life in a better way.

Q.2 Who Discovered cells?

Ans- Robert Hooke discovered the cell in 1665. Robert Hooke observed a piece of bottle cork under a compound microscope and noticed minuscule structures that reminded him of small rooms. Consequently, he, named these rooms as cells. However his compound microscope had limited magnification, and hence, he could not see any details in the structure. Owing to this limitation, Hooke concluded that these were non-living entities.

Later Anton Van Leeuwenhoek observed cells under another compound microscope with higher magnification. This time he had noted that the cells exhibited some form of movement. (motility). As a result, Leeuwenhoek concluded that these microscope entities were alive. Eventually, after a host of other observations these entities were named as animalcules. In 1883, Robert Brown, a Scottish botanist, provided the very first insights into the cell structure. He was able to described the nucleus present in the cell of orchids.

Characteristics of cells

Following are the various essential characteristics of cells:

1- Cell provide structure and support to the body of an organism.

2- The cell interior is organised into different individual organelles surrounded by a separate membrane.

3- The nucleus (major organelle) holds genetic information necessary for reproduction and Cell growth.

4- Every cell hat one nucleus and membrane bound organelles in the cytoplasm.

5- Mitochondria, a double membrane-bound organelle, is mainly responsible for the energy transactions vital for the survival of the cell.

6- Lysosomes digest unwanted materials in the cell.

7- Endoplasmic reticulum plays a significant role in the initial organisation of the cell by synthesizing selective molecules and Processing, directing and sorting them to their appropriate locations.

Types of cells

Cells are similar to factories with different labourers and departments that work towards a common objective. Various types of cells perform different functions. Based on cellular structure there are two types of cells.

1- Prokaryotes, 2- Eukaryotes

Prokaryotic cells

1- Prokaryotic cells have no nucleus. Instead some prokaryotes such as bacteria have a region within the cell where the genetic material is freely suspended. This region is called the nucleoid.

2- They all are single-celled microorganisms. Example include archaea, bacteria and cyanobacteria.

3- The cell size range from 0.1 to 0.5 μm in diameter.

4- The hereditary material can either be DNA or RNA.

5- Prokaryotes generally reproduce by binary fission, a form of asexual reproduction. They are also known to use conjugation- which is often seen as the Prokaryotic equivalent to sexual reproduction (however, it, is not sexual reproduction.)

Eukaryotic cells.

1- Eukaryotic cells are characterized by a true nucleus.

2- The Size of the cells ranges between 10-100 μm in diameter.

3- This broad category involves Plants, fungi Protozoans, and animals.

4- The plasma membrane is responsible for monitoring the transport of nutrients and electrolytes in and out of the cells. It is also responsible for cell to cell communication.

5- They reproduce sexually as well as asexually.

6- There are some contrasting features between plant and animal cells. For eg. - the plant cell contains chloroplast, central vacuoles, and other plastids, whereas the animal cells do not.

Cell Structure

The cell structure comprises individual components with specific functions essential to carry out life's processes. These components include- cell wall, cell membrane, cytoplasm nucleus, and cell organelles. Read on the to explore more insights on cell structure and functions.

Cell Membrane

1- The cell membrane supports and protects the cell. It controls the movement of substances in and out of the cells. It separates the cell from the external environment. The cell membrane is present in the cells.

2- The cell membrane is the outer covering of the cell within which all other organelles, such as the cytoplasm and nucleus, are enclosed. It is also referred to as the plasma membrane.

3- By Structure, it is a porous membrane (with Pores) which permits the movement of the selective substances in and out of the cell. Beside this, the cell membrane also protects the cellular component from damage and leakage.

4- It forms the wall-like structure between two cells as well as between the cell and its surroundings.

5- Plants are immobile, so this cell structures are well-adapted to protect them from extend factors. The cell wall helps to reinforce this function

Cell wall

1- The cell wall is the most prominent part of the plant's cell structure. It is made up of cellulose, hemicellulose and pectin.

2- The cell wall is present exclusively in plant cells. It protects the plasma membrane and other cellular components. The cell wall is also the outermost layer of plant cells.

3- It is a rigid and stiff structure surrounding the cell membrane.

4- It provides shape and support to the cells and protects them from mechanical shocks and injuries.

Cytoplasm

1- The cytoplasm is a thick, clear, jelly-like substance present inside the cell membrane.

2- Most of the chemical reactions within a cell take place in this cytoplasm.

3- The cell organelles such as endoplasmic reticulum, vacuoles, mitochondria, ribosomes, are suspended in this cytoplasm.

Nucleus

1- The nucleus contains the hereditary material of the cell, the DNA.

2- It sends signals to the cells to grow, mature divide and die

3- The nucleus is surrounded by the nuclear envelope that separates the DNA from the rest of the cell.

4- The nucleus protects the DNA and is an integral component of a plant's cell structure.

Cell Organelles

Cells are composed of various cell organelles that perform certain specific functions to carry out life's processes. The different cell organelles along with its principle functions are as following-

Cell organelles and their functions.

Nucleolus- The nucleolus is the site of ribosome synthesis. Also, it is involved in controlling cellular activities and cellular reproduction

Nuclear membrane- The nuclear membrane protects the nucleus by forming a boundary between the nucleus and other cell organelles.

Chromosomes- Chromosomes play a crucial role in determining the sex of an individual. Each human cells contain 23 pairs of chromosomes.

Endoplasmic reticulum- The endoplasmic reticulum is involved in the transportation of substances through the cell. It plays a primary role in the metabolism of carbohydrates, synthesis of lipids, steroids and proteins.

Golgi bodies- Golgi bodies are called the cell's post office as it is involved in the transportation of materials within the cell.

Ribosome- Ribosome are the protein synthesizers of the cell.

Mitochondria- The mitochondria is called the powerhouse of the cell. It is called so because it produces ATP-the cell's energy currency

Lysosomes- Lysosomes protect the cell by engulfing the foreign bodies entering the cell and help in cell-renewal. Therefore, they are known as the cell's suicide bags.

Chloroplast- Chloroplast are the primary organelles for photosynthesis. It contain the pigment called chlorophyll.

Vacuoles- Vacuoles store food, water and other waste materials in the cell.

Cell theory

Cell theory was proposed by the German Scientists, Theodor Schwann, Matthias Schleiden and Rudolf Virchow. The cell theory starts that:

i- All living species on earth are composed of cells

ii- A cell is the basic unit of life

iii- All cells arise from pre-existing cells.

A modern version of the cell theory was eventually formulated, and it contains the following postulates.

i- Energy flows within the cells

ii- Genetic information is passed on from one cell to the other.

iii- The chemical composition of the cells is the same.

Functions of cells.

A cell performs major functions essential for the growth and development of an organism important functions of cell are as following provides support and structure.

All the organisms are made up of cells. They form the structural basic of all organisms. The cell wall that function to provide support and structure to the organisms. For eg.-the Skin is made up of large no. of cells. Large numbers of cells. Xylem present in the vascular plants is made up of cells that provide structural support to the plants.

Facilitate growth Mitosis

In the process of mitosis, the parent cell divides into the daughter cells. Thus the cells multiply and facilitate the growth in an organism.

Allows transport of substances.

Various nutrients are important by the cells to carry out various chemical processes going on inside the cells. The waste produced by the chemical processes is eliminated from the cells by active and passive transport. Small molecules such as oxygen, Carbon dioxide and ethanol diffuse across the cell membrane along the concentration gradient. This is known as passive transport. The larger molecules diffuse across the cell membrane through active transport where the cells require a lot of energy to transport the substances.

Energy Production

Cells require energy to carry out various chemical processes. This energy is produced by the cells through a process called photosynthesis in plants and respiration in animals.

Aids in Reproduction.

A cell aids in reproduction through the processes called mitosis and meiosis. Mitosis is termed as the asexual reproduction where the present cell divides to form daughter cells meiosis causes the daughter cells to genetically different from the parent cells.

Thus, we can understand why cells are known as the structural and functional unit of life. This is because they are responsible for providing structure to the organisms and Perform several functions necessary for carrying out life's processes.

Function of Nucleus

i- It control all the cellular activity

ii- It store Genetic material

iii- It help in Protection synthesis

iv- It help in cell Division

v- It help in synthesis of RNA

vi- It is also called control body of cell Golgi body

Discovered by- Camillo Golgi in 1898 It is a membranous bound organelles that help in processing and packaging of protein Which formed by E.R- Edoplasic Reticulum

Edoplasmic Reticulum- It is the network membrain inside the cell through which proteins and other molecular move. It is of two type (i) Rough E.R- and (ii) smooth E.R

i- **Rough E.R-** Ribosomes molecules attached on the other surface. It help main synthesis of Proteins.

ii- **Smooth E.R**- Ribosomes molecules do not attached on the outer surface.

It helps main synthesis of non-protein substance like-steroid. Harmones, cholesterol. It also responsible for detoxification of toxic Substances.

Tissue

Q.1 What is Human Body Tissue?

Ans- Tissue in the Human body-Body tissue is composed of grouping of cells that make up out our organs and other body part. The Human body has four type of tissue.

1- Muscle tissue is found all throughout the body, even in the organs such as the heart

2- Epithelial tissue covers and protects our bodies and the lining of some organs in the form of skin.

3- Connective tissue, which can be found filling the spaces in our body, holds our parts together and provides us support

4- Nervous tissue transmits signals from nerves to the spinal cord and brain allowing us to use our five 5 senses.

Q.2 What is Body Tissue?

Ans- If you were to try to explain to someone what yours body is made of, our might say to arms, two legs, feet and hands a head and a torso. Or, you might go to the other extreme and say that our are made up of billions of cells. Bothe answers should be correct. However, there is a more specific way to described what made up a body. We are composed of several different types of tissue. But what exactly does that mean? Human body tissue- is another way of describing now our cells are grouped together in a highly organized manner according to specific structure and function. These grouping of cells form tissue structure, which them makeup organs and various parts of the body. For example muscles is easy to see and feel in the body, as it is one of the types of tissues.

There are four types of Tissue

We have determined that we are made up of four types of tissues. In addition to muscle tissue, we have connective epithelial and nervous tissues in the body. So what are these tissues is made up of and how are they different from one another? Let's zoom in an each one to better understanding.

Muscle Tissue- As mentioned carries, human tissue is made up of particular kinds of cells that work together. First let's look at muscles tissue. Muscle tissue is made up of excitable cells that are long and fibrous. There cells are ready for contraction, or the activation of tension in our muscles, making it possible for us to move our body parts. They are arranged in parallel lines and are bundled, making muscle tissue very strong. If you take a pile of rubber bands, line the up next to each other and attempt to stretch them you may get the idea of the nature of the muscle tissue.

Epithelial Tissue- Epithelial tissue is made up of epithelial cells, which are vastly different from the muscle cells we just talked about. These Cells can be flat, cuboidal, or columnar. They are joined tightly together, making a single or stacked continuous sheet. Like a quilt that is tightly stitched, epithelium makes an excellent protective cover for the body, in the form of skin. Epithelial tissue can also be found lining some internal cavities and organs.

Connective Tissue- As its name suggests, connective tissue makes up a connective web inside our body. Holding our body parts together and providing support are the main jobs of this tissue. We would certainly not be in good shape if all of our internal body parts were free-floating. Connective tissues fills in the spaces inside our body with a matrix made of fibers within a liquid, solid or jelly-like substance. Think of a gelatin salad with fruits suspended inside, and you will have an idea of how certain types of connective tissues function.

Nervous Tissue- Nervous tissue is found within the nervous system and is made up of unique specialized cells. Like electrical circuits, the nervous system transmits signals from nerves to the spinal cord and brain. Cells known as neurons conduct these impulses, making it possible for us to use our senses.

Examples of Tissues in the Human Body you already know several examples of tissues found in the human body. Muscle tissue is found in all of the obvious places, as in our biceps, triceps, quadriceps and so on. But muscle is also located in many of our internal organ such as the walls of our arteries and digestive tract. And don't forget about the most important muscle of all: the heart, cardiac tissue is also muscle tissue, as this powerful organ is constantly contracting to pump blood through out our body.

Epithelial tissue, as previously mentioned, is found both covering our body and lining some organs for example-our outer layer of skin is made up of epithelial tissue. This sheet of epithelial tissue is like a permanent waterproof coat for our body. It protects us from potential invaders like viruses and keep our body from loosing moisture. The lining of the mouth and esophagus are also example of epithelial tissue

Connective tissues comes in several different forms. The most abundant in the body is loose connective tissue, and it is found filling the space in our body, such as in our collagen blood, cartilage, bones and several other areas. For example-If you pull on your skin it will stretch but only so far. The connective tissue underneath it keeps it attached so that our skin isn't flopping around. Fat tissue is another example of connective tissue, as it provides casboining and support.

Finally, nervous tissue is clearly found within the nervous system. Neurons extend throughout the body making it possible for us to use our 5 five senses. The brain and spinal cord make up the rest of this type of tissues.

Tissue- It is defined as group of similar cells that have similar structure and that function as a unit.

1- **Epitherial Tissue-** It is mostly present on body surface (epidermis) and lined hollow organs, body cavity, glands and ducts.

2- **Epithelium Tissue-** It is available but has their own serve supply No blood circulation happen in it. It is divided into two main classification.

i- **Simple Epithelium Tissue-** It consist of single layer of cell. It is classified into 4 Four group.

(1) **Squarnous Epithelium Tissue-** It consist of single layer of flat cells. It is mainly present in alveoli, bowmen's capsule.

(2) **Cuboidal Epithelium Tissue-** It consist of cube shape cells. It is mainly present in real tubules lining of ducts of glands.

(3) **Columnar Epithelium Tissue-** it consist of rectangular shape cells. It is mainly present in lining of GII, ducts of many glands Gastro Internal Track

(4) **Ciliated Epithelium Tissue-** It consist of ciliated rectangular shape cells It is mainly Present in respiratory tract, uterus fallopian tube

ii- **Stratified Epithelium tissue-** It consist of several layers of different shape cell. It protect the underlying structure from wearing and tearing. It is divided into two main classification

1- **Stratified squamous Epithelium tissue-** Transitional epithelium tissue- It consist of flat cells start it again classified into two type

a- Keratinized E.T- It found on dry surface eg.- kin, Hair, Nails.

b- Non Keratinized E.T- It found on wet surface eg- oral cavity, anal caneal, vagina

2- **Transitional epithelium Tissue-** It consist of several shape's of cells. It is found only in urinary system.

Connective Tissue- Tissue that supports and connects other tissues and parts of the body. It is most abundant and widely distributed tissue in our body. Connective tissue is classified into two types.

A- Loose Connective Tissue- It contain less fibers and more cells closely intertwined So cells and fibers are loosely arranged It again divided into following subtype.

1- **Areolar connective Tissue-** More widely distributed in the body, present into almost every structure of body so called packing material of body. It consist of fibroblasts collagen and elastin fibers.

2- **Adipose connective Tissue-** It consists of adipose cell which filled with single droplet of trigly ceride in their cytoplasm. Function- Reduce heat loss, serve energy reserve and support and protect organ.

B- Dense connective Tissue- It contain more number of fibers so cells and fibers are compactly packed eg.- tendons and ligaments.

Functions of connective tissues.

1- Binds together, support and strengthens other body tissue

2- Insulate and protect internal organ eg- bones.

3- Storage like adipose tissue (store fat)

4- Main site of immune reaction. Like kupffer cells and macrophage.

Muscular Tissue- Muscular tissue is composed of cells that have the special ability to shorten or contract in order to produce movement of the body parts.

Classification of Muscular Tissue

1- **Skeletal Muscle Tissue-** skeletal muscle fibers are cly, cylindrical, multinucleated and under voluntary control.

2- **Smooth muscles Tissue-** It is also called involuntary muscle. It consist of narrow spindle-shaped cells.

3- **Cardiac Muscle Tissue-** Cardiac muscle or myocardium makes up the thick middle layer of the heart. This is a specialized muscles this is present only in heart. In myocardium ability nerve impulses through own conducted help in heart contract or relax by cardiac muscles.

Muscular tissue- what make muscular tissue unique from the other three types.

Contract (shorten in length) When stimulated by an electrical impulse. What is the electrical impulse comes from? The electrical impulse comes from the nervous system.

Basic structure of muscle tissue

Muscle cells are very in shape

Long cylindrical Fiber, Rectangular & Branches spindle shape. Muscle cells can Have one nuclei per cells: uni-nucleated and the last type of Nuclie Cell have more than oen nuclei per cell: Called Multi-nucleated.

Muscular Tissue

Control of muscle

Voluntary Control-Under conscious control you control it by thinking about it.

Involuntary control- non under conscious control- you can-not control it by thinking about the structure of skeletal muscle it is under voluntary control and it is attached with the bones and skeletons System when skeleton muscles controls it pope on the bones and movement of the joints and it function contracts to perform movements

Structure- Long cylindrical, multinucleated (Many nuclei per cell) striated alternating light and dark bands.

The second muscle tissue is cardiac muscle it is involuntary control and located in heart and its function is to contracts to pump blood its structure and circulate blood from heart to the body is rectangular branched uni-nucleated cone nucleus per cell, striated.

Smooth Muscles- It is involuntary control it is located in Blood vessels air passage ways. Digestive system, passage ways, hollow organs eyes. Its function is to contracts to constrict vessels or passage ways movement of food, movement of blood its structure is spindle shaped (tapered ends) uni- nucleated non-started.

Q.3 What makes Nervous tissues unique the other three types of tissues?

Ans- Nervous tissues has the ability to produce an action potential the communicate nervous tissue has the ability to Create, receive and conduct electrical impulses coordinates and controls body activities.

Neuron- It is the nerve cell of the nervous system it receives, creates and conducts action potential nerve impulses. It has main 3 parts.

1- Dendrite- Responsible for receivers electrical signals

2- Cells body- contains organelles like the nucleus

3- Axon-transmits electrical signals Neurologlial cell- The primary function is to supports, nourishes and maintains the health of the neuron

There are several types

1. Microlia- the microglia is Macrophage which is engulfs foreign substance.

2. Astrocytes- is provides nutrients to Nucleus.

3. Oligodendrocytes- forms mylin sheaths in central nervous system.

4. Schwan cells- Form myelin sheaths in peripheral nervous system.

5. Ependymal cells-Line ventricles of the brain and central canal of 5 pinal cord. Or we can say it is like a CPU- Central processing Unit of our body from where whole body is controls.

Muscles

Q.1 What are the Muscles?

Ans- Muscles are soft tissues fond in most animals and humans both. The cells of muscles comprise protein filaments of action and myosin that slide past one another, which produces contraction and changes both the length and the shape of the cell.

The term muscle is derived from the Latin word "Muscles" which refers to a little mouse which is due to the shape of certain muscles or the contraction of muscles that look like a moving mouse.

In humans, muscles function by producing force and motion and are primarily responsible for:

1. Locomotion

2. Maintaining and changing body posture.

3. Circulation of blood cells throughout the body

4. Movement of internal organs, such as the contraction of the heart and the movement of the food through the digestive system via peristalsis.

The human muscles:- The human muscular system includes more that 600 muscles, which make up about 40 to 50 percent of the total body weight. These muscles are attached to bones, blood vessels and other internal organs of our body and are mainly compose of skeletal muscles, tissue, tendons and nerves. The muscles of the human muscular system are composed of a kind of elastic tissue.

Every movement in our body is the result of muscles. Contraction and is found in every organ, including the blood vessels, heart, digestive organs etc. In this- In these organs muscles function by transferring substances throughout the body. There are three types of muscle and are mainly classified based on their movements and structures. The energy required for the functioning of muscles is Predominantly Powered by the oxidation of fats and carbohydrates particularly and from the stored energy molecules adenosine triphosphate (ATP)

Types of Muscules

Skeletal muscles- skeletal muscle is a muscle tissue that is attached to the bones and is involved in the functioning of different parts of the body. These muscles are called voluntary muscles as they come under the control of the central nervous system in the body.

Structure of Skeletal Muscle- skeletal muscle is a series of muscle fibers composed of muscle cells, Which are long and multinucleated.

Skeletal muscles are cylindrically shaped with branched cells attached to the bones by an electric tissue or collagen fibers called tendons. Which are composed of connective tissues.

The end of each skeletal muscle has a tendon, which connects the muscle to bone and connects directly to the collagenous, the outer covering of Skeletal muscle.

There is a group of muscle fibers, present below the epimysium, which are collectively called the fascicles. These muscle fibers are surrounded by another protective shield formed from collagen.

The Perimysium, a sheath of connective tissue surrounding the muscle fibers allows nerve and blood vessels to make their way through the muscle.

Functions of skeletal Muscle

1. It maintains body posture
2. It regulate body temperature
3. It connects to and controls the movements of the skeleton.
4. It is responsible for performing muscular involuntary movements.
5. It is responsible for body movements such as breathing, extending the arm, typing writing etc.
6. It is responsible for the erect posture of the body. The Sartorius muscles in the thighs are responsible for body movement.
7. The skeletal muscles protects the internal organs and tissues from any injury and also provide support to these delicate organs and tissues.
8. These also support the entry and exit points of the body.

For eg- the sphincter muscles present

1- **Around the mouth-** These muscles reduce the size of the openings by the contraction of the muscles and facilitate the swallowing of food.

2- **Around the urinary tract-** These muscles control urination by the contraction of the muscles in the urithra.

3- **Around the anus-** These muscles reduce the size of the openings by the contraction of the muscles and facilitate defecation.

2- **Cardiac muscles-** Cardiac muscles are found only in the heart and are self-stimulating, which has an intermediate speed of contraction and energy requirement. This muscle is not port of the musculoskeletal system.

Cardiac muscles are striated muscles, which are responsible for keeping our heart functioning by pumping and circulating blood through out the body and performing muscular involuntary movements, they are involved in continuous rhythmic contraction and relaxation. The interconnected muscle cells or fibers provide strength and flexibility to the cardiac muscle tissue.

3- Structure of a cardiac muscle cardiac muscle exists only within the human heart. It is a specialized form of muscle evolved to continuously and repeatedly contract, providing circulation of blood throughout the body. Cardiac muscle has a regular pattern of fibers similar to that of smooth muscles. These muscles comprise the cylindrical, branched fibers and centrally located nucleus. The T-Tubules or transverse tubules are rich in iron channels and are found in the atrial muscle cells.

These muscles are striated muscles with cylindrical-shaped cells, which include intercalated discs and Join neighbouring fibers.

Functions of a cardiac muscles

The primary function of the cardiac muscle is to regulate the functioning of the heart by the relaxation and contraction of the heart muscles other functions of cardiac muscles include.

1- The cardiac muscles function as the involuntary muscle.

2- The cardiac muscles are also involved in the movement or the locomotion.

3- The cardiac muscles work without stopping day and night. They work automatically and make the heart contract so that the heart can squeeze the blood vessels and release them so that the heart can fill up with blood again

4- The heart comprises a specialized type of cardiac tissue, which consists of "Pacemaker" cells. These contract and expand a in response to electrical impulses from the nervous system

4- SMOOTH Muscles.

Smooth muscles are non-striated, involuntary muscles, Which are controlled by the autonomous Nervous System (ANS) These muscles are found almost in all organ such as the stomach, bladder, blood vessels, bike ducts, the eye, the sphincters the uterus etc.

The smooth muscles function by stimulating the contractility of the digestive, urinary, reproductive systems, blood vessels and airways. These smooth muscles are spindle-shaped with a single nucleus. They are under involuntary control therefore you cannot move these muscles with conscious throught.

Structure of smooth muscles

The smooth muscles of the human muscular system are spindle-shaped muscle fibers with a single nucleus. The thickness of the smooth muscles ranges between 3-10 μm and their length ranges between 20 to 200 μm, which are shorter compared to the skeletal muscle. These muscles lack filaments, special protein, action and myosin and produce their own connective tissue functions of a cardiac muscle.

Like all other types of muscles, smooth muscles are also involved in contraction and relaxation. Other functions of smooth muscles include:-

1. It is involved in the sealing of orifices.
2. It produces connective tissue proteins such as collagen and elastin
3. Transports chyme (a pulpy acidic fluid) for the contractions of the intestinal tube.
4. Smooth muscle plays a vital role in the circulatory system by maintaining and controlling the blood pressure and flow of the oxygen through out the body.
5. Smooth muscles are also responsible for

a. Contracting the irises.

b. Raising the small hair on your arm

c. Contracting the spincters in our body

d. In the movement of the fluids through organs

e. It is much more useful for providing consistent and elastic tension.

Based on the muscle action, muscles are further classified into.

Voluntary muscles- Voluntary muscles are long, multinucleated cells, containing sarcomeres arranged into bungles. These muscles are composed of cylindrical fibers and are usually attached to bones and the skin. They play an important role in allowing the body to move by contracting by and relaxing and their actions are mainly under the control of the somatosensory nervous system. These voluntary muscles include skeletal muscles.

Involuntary muscles

Involuntary muscles are striated and branched in the case of cardiac muscle. The actions of involuntary muscles are mainly controlled by the autonomic nervous system in the body. These involuntary muscles include smooth muscles and cardiac muscles.

The muscular system is an organ system consisting of skeletal, smooth and cardiac muscle. It permits movement of the body, maintains posture, and circulates blood throughout the body. The muscular systems in vertebrates are controlled through the nervous system although some muscles (such as cardiac muscle) can be completely autonomous. Together with the skeletal system in the human. It forms the muscle skeletal system, which is responsible for the movement of the body.

Bone

Q.1 What is Bone-Defination?

Ans- The bone is rigid body tissue that makes up our body skeleton. The bone is a connective tissue that is made up of different types of cells. Internally, it has a honey comb-like matrix that gives rigidity to bones. The primary function of the bones is to provide structural support to the body and enable mobility.

They also produce RBCs and WBCs and store minirals inside them.

Bone Structure and Morphology

The bone is made up of varied composition, that is about 30% flexible matrix and 70% bound minerals that are embedded with specialized bone cell. This unique constitution of the bone allows it to be strong and hard while being lightweight. The bone matrix is compared of collagen (90-95%) and ground substances. The collagen fibers provide elasticity and resistance to the structure. The matrix allow also contains calcium phosphate that hardens the bone structure.

The bones have an external layer called the cortex. It is a hard exterior that gives the white and smooth appearance to the bones. It forms around 80% of the total mass of the human skeleton. The cortical surface of the bone is covered by the periosteum on the outside and by endosteum on the inside. The interior of the bone is filled with a spongy Tissue that is referred to as cancellous bone or trabeculae. It is an open porous network of intersecting plates and spicules that gives it a higher surface area to volume ratio. It is a vascular tissue that often contains red bone marrow and is responsible for hematopoiesis. Bone marrow or mycloid tissue is found in the cancellous tissue in newborns that produce red blood cells. As the child grows, the red bone marrow is converted to adipose tissue, that is a fatty substance cells of the Bone

The bone consists of four types of cells which have different functions. Let us look at them briefly.

1. Osteoblasts- Osteoblasts are large cuboidal cells that form about 4-6% of the total bone cells. These cells synthesis and mineralize bone during bone synthesis and remodeling. They form a closely packed sheet like surface on the bone. They are formed by the differentiation of osteogenic cells. It produces many cellular products such as alkaline, phosphatase, growth factors, collagenase and collagen fibers.

The osteoblasts get surrounded by the bone matrix and cellular products and are then referred to as osteocytes.

2. Osteocyte- osteocytes are oblate shaped cells that are derived from osteogemic cells or osteoblasts. They constitute almost 95% of the total bone cells. They are located in small spaces in the bone matrix called lacunae. The cytoplasmic Processes of the osteocytes extend towards other osteocytes in small channels called canaliculi. Through these channels, the nutrients and waste products are exchanged that helps in maintaining the viability of osteocytes.

They are the most abundant and long living cells of the bone. They also take part in bone deposition, resorption and remodeling.

3. Osteoclast- Osteoclasts are large multinucleated cells that break down bone tissues. Bones are dynamic tissues that undergo constant remodeling to put up with the stress and requirements of our body osteoclasts are phagocytic cells that are derived from the macrophage-monocyte cell lineage.

The major function of these cells is bone resorption. Bone resorption is the process of breaking down of bone tissues by the osteoclasts to release minerals such as calcium into blood circulation.

Osteoclasts acieve the destruction of bone tissues by secrating certain enzymes such as acid phosphatase. This enzyme is capable of digesting collagen as well as calcium and phosphorus. The bone tissues are first broken down into fragments that are later engulfed by the osteoclasts and digested within the cytoplasmic vacuoles. The minerals released from the digestion are released into the bloodstream.

4. Osteogenic Cells- osteogenic cells or osteoprogenitor cells are stem cells present in the bone. These are precursor cells that give rise to specialized cells, viz, osteocytes and osteoblasts. These cells reside in the bone marrow.

Osteogenic cells are proliferative cells that are derived from the mesenchymal stem cells. Morphologically, They appear like a flattened, spindle shaped cell.

Types of Bone

There are five types of bones found in our body. Let us liscous them one by one.

1. Long Bones- Long bones are bones that are longer then they are wide. The mid section of the long bone is called diaphysis that is Predominatly made up of cortical bone and also contains bone marrow and adipose tissue. The diaphysis flares and form an internal cancellous structure called metaphysis. The epiphysis is the rounded end of the long bones (on both sides) that has an internal cancellous structure.

Examples of long bones in humans include the femur, tibia, and fibula of the legs, humerus, radius and ulna of the arm, clavicle (collar bone) metatarsals and metacarpals of the feet and hand respectively.

2. Short Bones- Short bones are the ones that are as long as they are wide. They are cube shaped bones that have a thin cortical layer and a thick spongy inferior. They provide stability and support to our body. Examples of short bones include tarsals and carpals in the foot and band, respectively.

3. Flat Bones- Flat bones are thin and curved bones that are composed of spongy cancellous tissue sandwiched between two thin layers of cortical bone. The main function of flat bone is to provide a broader surface area for the attachment of muscles. They usually form broad flat plates as in the sternum, cranium (Skull) rib cage and ilium (pelvis)

4. Sesamoid Bones- Sesamoid bones are the bones that are embedded in the tendons of muscles. The patella of the knee and pisiform of the wrist are two examples of sesamoid bones.

5. Irregular Bones- Irregular bones, as the name suggests, do not fit into any of the bone types and form peculiar bony shapes in the human endoskeleton. They are composed of cancellous tissue that is enclosed between a thin layer of cortical bone. Example of the irregular bones include hyoid, sacrum, coccyx, maxilla mandible etc.

Functions of the Bone

1- Mechanical Functions

(i) The bones provide a framework for the human body. They keep the body supported and form attachment points for ligaments, tendons, Joints, and skeletal muscles.

(ii) They protects the internal organs such as the skill that protects the brain and ribs that protect the heart and lungs.

(iii) The three 3 ear bones namely malleus, incus stapes help in the process of hearing by sound transduction.

Synthetic Functions

(i) The bones consist of bone marrow that carries out the process of hematopoiesis and produces red blood cells, white blood cells and platelets.

(ii) The bone marrow is also the site where defective red blood cells are destroyed

Metabolic Functions

(i) The bone is a mineral reservoir for the storage of calcium and phosphate.

(ii) The adipose tissue in the bone marrow serves as a site for storage of fatty acids.

(iii) Bone releases alkaline salts into the blood stream that maintain an acid-base balance.

(iv) The bones absorbs heavy metals and other foreign elements from the tissue to reduce their negative effect and later excreates them.

(v) It releases the enzymes osteocalcin that regulate blood shuger sugar levels and fat deposition.

(vi) Bone resorption- the process of release of calcium in the bloodstream is significant bone function that maintains the calcium balance in our body.

Bone Remodelling and Resorption.

The process of bone formation is called ossification. The process of continuous creation and replacement of bone cell is known as bone remodeling, osteoblasts and osteoclasts accomplish the process of remodeling by a cascade of signals. The remodeling process is vigorous till the age of 35 after which it starts to decrease. Bone remodeling is important for repair of damaged bones, maintenance of calcium homeostasis and shaping the skeleton during growth years.

The process of remodeling is achieved through bone resorption. Bone resorption is the breakdown of osteoclasts to release the minerals such as calcium from the bone tissue to the blood bone formation is higher that resorption during childhood, but resorption exceeds the process of formation as ageing begins.

Bones provides the structure of our body and Bones hold the body together adult human skeletons made up of 206 bones or in babies around 300 bones. Bones material are strong, samirijid bones is a durable connective tissues made up of coligen & calcium and some bones cells.

Functions of Bones

Bones is a rigid tissue that make up the skeleton and providing body from work and give attachments to muscles to allow mobility. Bones provides protection to organs and give structure and support to our body

Broadly Five function of Bones

1- They give support to our body

2- It provides levers for movement

3- Bones provides protection to our internal organs.

4- It is house of Bone marrow. Bone marrow is the spongy substance inside the bone it is present in center of bone. It manufactures blood cells called-Red blood cells (RBC) White blood cells (WBC) and platelets. RBC There is three type of Blood cells. RBC Carries oxygen. WBC kills micro organisms and platelets help in blood cloting.

5- Phosphorus- It is a key energy of bones phosphorus is evolved in energy transformation machmanism including the Generation of ATP – Atrocin try Phosphate.

It stores calcium and phosphorus. Calcium is a major component of bones. Our body is not able to form calcium. We take calcium from outside products like milk products egg etc from these food items our body absorbs calcium so we understand why calcium is important for our body.

It is important for muscles contraction make bones strong and healthy.

A Bone is a rigid organ that constitutes parts of the skeleton in most vertible animals Bones protects the various other organs of the body, produce RBC and WBC and platelets, and store minerals, provide structure and support for the body and enable mobility Bones come in variety of shape and size and have a complex internal and external structure. They are light weighted yet strong and hard and serve multiple function.

Bone Tissue- Cosseous tissue, which is called bone in the uncountable sense of the work word is hard tissue, a type of specialized connective tissue, it has a honey comb like matrix internally which help to give the bone rigidity. Bone tissue is made of different types of Bone cells. Osteoplasts and osteocytes are involved in the formation and mineralization of bones osteoclasts are involved in the resorption of bone tissue. Modified flattend osteoblasts become the lining cells that form a protective layer on the bone structure. The mineralized matrix of bone tissue has a organic component of mainly collage called ossein and an inorganic component of bone minerals made up of various salts. Bone tissues is mineralized tissues of two type, cortical bone and cancellous bones other types of tissues found in bones including bone marrow endosteum, Periosteum nerves blood vessels and cartilage. In the human body at birth there are approximately 300 bones present many of these fuse together during development, leaving a total of 206 separate bones in the adult, not counting numerous small seasamoid bones the large bone in the body is the femur or thigh bone, and the smallest is the stapers in the middle ear joint. In this Joint one articular surface is spherical (Ball like) other articular surface is cup like cavity eg- shoulder, hip-joint

2- Hinge Joint- In this joint movement take place in one place only movement flexion and extension eg- elbow, knee-joint.

3- Pivot Joint- In this type of Joint a rounded process of bone rotates within ring Movement only rotation eg-Radio-ulnar Joint

4- Condyloid/Ellipsoid Joint- One articular surface is concave another is convex.

Movement-flexon, extension, Adduction abduction eg- wrist Joint.

5- Saddle/sellar Joint- Articular surface is concave-convex (saddle shape) Movement- All places eg-Carpometacarpal Joint.

6- Plane Joint- In this type of Joint both articular surface is plane/flat eg-acromio-Clavicular Joint.

Human skeletal system

Skeleton System is a system which present in our body it provide rigidity to our body it provide base to our body and it also give size and defenate shape to our body so it's a very important system to our body which protect our soft organs such as heart lungs which store calcium, phosphorus and also it help the formation of (RBC) as we know that bone mero forms (RBC) and the bone merrow is present in side the bone so the skeleton system is helps in formation of RBC Cartilage. There are 206 bones is present in our body.

Joints

Joints- A Junction at which two or more bones are articulates with each other structural classification of joints.

1- Fibrous Joint- It is the type of Joint that articulating surface is connected by fibrous tissue.

No movement can be possible

Eg- sutures on skull

2- Syndesmosis- Csyndesmos- Ligaments In this joint bones are united Joints by the fibrous membrain or tissue (Interosseous membrain)

3- Gomphosis- It is the special type of joint between tooth and socket

2- Cartilaginous Joints- Primary/ secondary primary cartilagios/ hyaline cartilaginous joint. The bone are united by hyaline cartilage. It permits slight movement during early life.

Secondary cartilaginous- It is called fibrocartiliginous Joint. Bone are united by strong fibrous tissue or fibrocartilage.

Intervertical Joint

3- Synoval Joint- Most common Joint present in air body, Most movable, freely movable joint of body-They are called synoval joints because they are lined with a synoval membrain and they connect. They contain lubricating fluid called synoval fluid.

Synovial membrain is a serious membrain.

They are 6 six type

1- Ball and socket joints- This is freely movable we understand this by human skeletal system of human body.

Osteology Arthrology	Human skeletal Division of Bones
1- Skull 29 Bones	Apendicular skeleton
• Cranium 08 Bones	1- Upper Limb
• Face 14 Bones	• Pectoral Girdle -04 Bones
• Ear 06 Bones	• Hand Bones – o6 Bones
• Hyoid 01 Bones	

2- Vertebral Calumn -26 Bones 3- Sternum – 1 Bone 4- Ribs -24 Bone Total 80 Bone	2- Lower Limb • Pelvic Girdle – 02 Bones • Leg Bones – 60 Bones

Q.1 What is a Joint?

Ans- A Joint generally means a point where two or more things are connected together. In this scenario, it is the point where two bones intersect. Joint means an articulation or in other words, a strong connection that joins the bones, teeth and cartilage together. It is necessary for all types of movement in the body involving bones. The force generated by through various Joints . The degree and ease of movement at different Joints vary to a lot of different factors. They could be classified based on two different things.

Classification of Joints

A Joint is classified based on functionality as to how much movement it allows:

1- Immovable: A Joint which permits no kind of movements is called synarthrosis. The sutures of the skull and also the gomphosis connecting teeth to that of the skull are some example of synarthrosis.

2- Shightly movable: An amphiarthrosis usually allows a very little amount of movement at one of the joints. Example of this amphiasthrosis include some of the intervertebral disks present in the spine and also the public symphysis present in the hips.

3- Freely movable: The third class of functional Joints is said to be freely moving kind of diarthrosis joints. Diarthrosis are said to have the highest range of motion of any kind of joint and also includes the knee, elbow, shoulder and also wrist

Joints are also classified based on the structure of the material present in them.

4- (i) A fabrus Joint is made up of tough collagen fibre, which does not allow any type of movement and also includes the structures of the skull along with the syndesmosis holding together the ulna in addition to the radius of the forearm.

(ii)Cartilaginous Joints- are said to be made up of bones that are Joined together with the help of cartilage. Some example include the joints present between the ribs and the costal cartilage and also the intervertebral discs present in the spine of the body.

(iii) The most common kind of Joint, the synoval joint is said to be characterized by the presence of fluid-filled space between the smooth cartilage pads present at the ends of the articulating bones. Such an arrangement allows movement. The synoval membrane that lines up the capsule produces oily snynovial fluid and lubricates joints, reducing friction and wear & tear

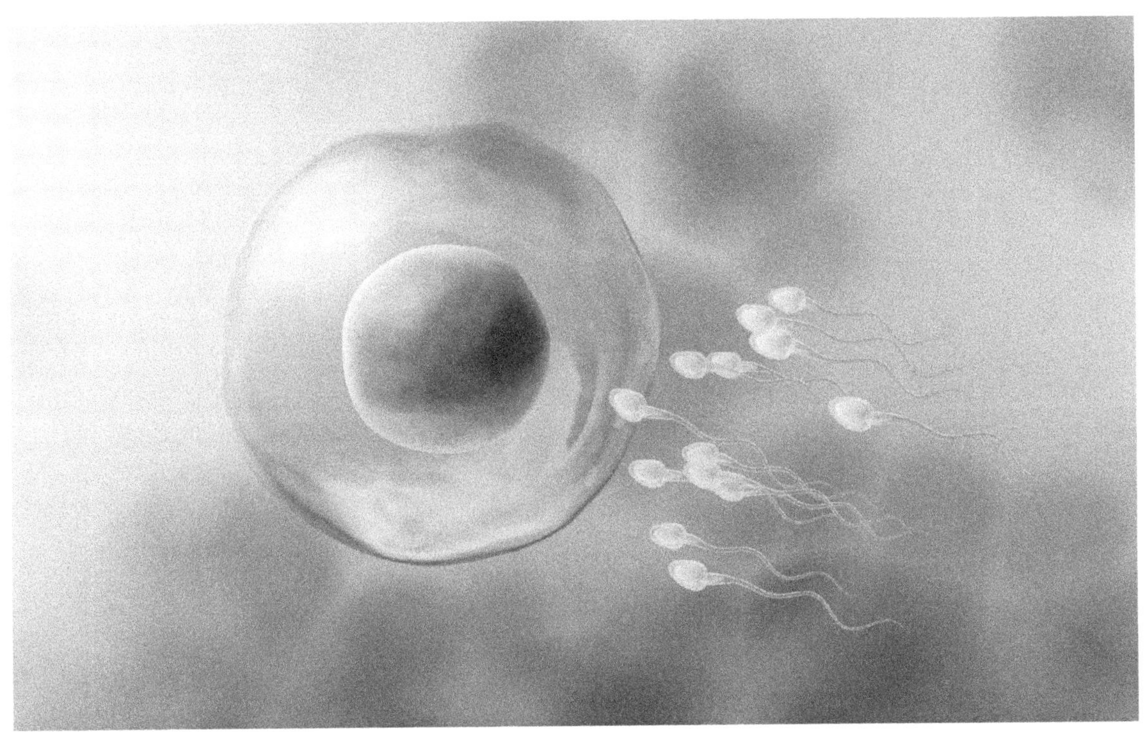

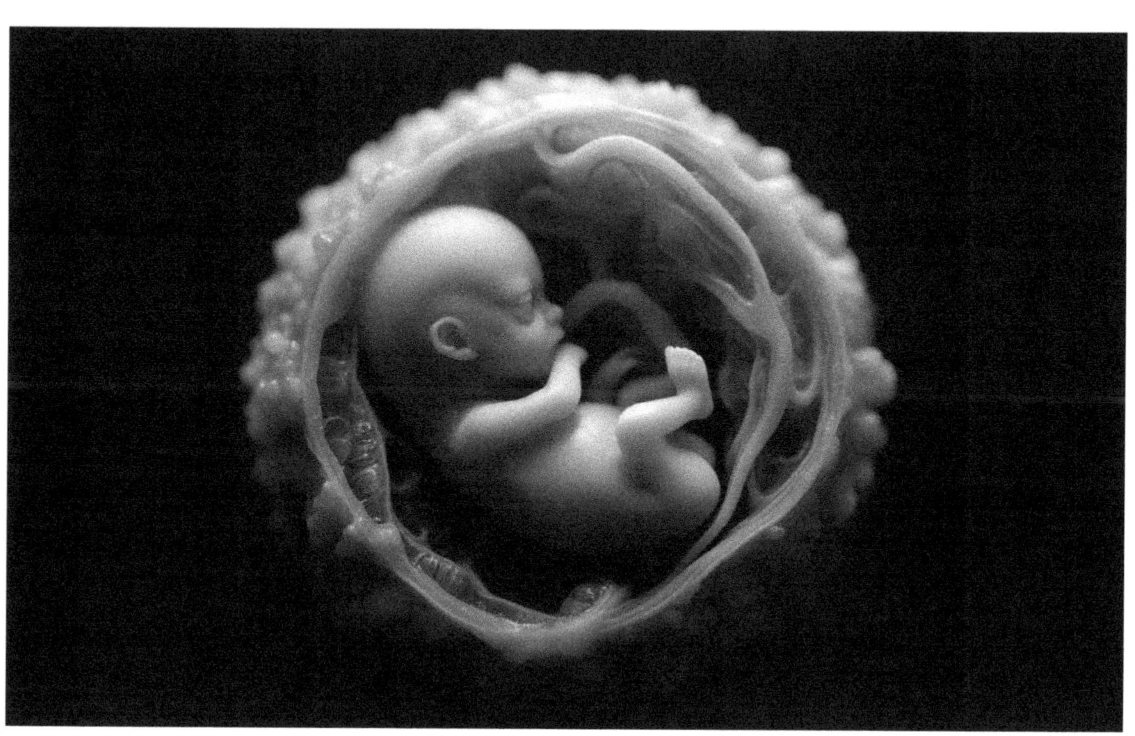

Chapter 16
Reproduction

Q.1 What is Reproduction?

Ans- Reproduction is a biological process by which an organsism reproduces an off spring that is biologically similar to the organism. Reproduction enables and ensures the continuity of species generation after generation. It is the main feature of life on earth.

Let su us have a detailed overview of reproduction, its types and the modes of reproduction in plants and animals.

Types of Reproductions

Asexual Reproduction- Asexual reproduction refers to the type of reproduction in which only a single organism gives rise to a new individual

Asexual reproduction does not involve the fusion of gametes, and therefore, the offsprings produced are genetically identical to the parent. The organisms produced by asexual reproduction are less diverse in nature. This type of reproduction is practiced widely by unicellular organisms.

The process involves rapid population growth and no mate is required for the process. However, a lack of genetic diversity makes organisms more susceptible to diseases and nutrition deficiencies.

Asexual reproduction is further divided into.

1- Binary fission- In this, the cell splits into two each cell carrying a copy of the DNA from the parent cell eg- amoeba.

2- Budding- In this, a small bud-like outgrowth gives rise to a new individual. The outgrowth remains attached to the organism until it is full grown. It detaches itself and lives as an individual organism. For eg- hydra

3- Fragmentation- In this the parent organisms splits into several parts and each part grows into a new individual for eg- planasia

4- Sporogenesis- In this type of reproduction a new organism grows from the sporse.

These can be created without fertilization and can spread through wind and animals.

Sexual Reproduction

Sexual reproduction is a type of reproduction that involves the production of an offspring by the fusion of male and female gametes. In the sexual reproduction male and female gametes are formed to produce an offspring these gametes are either formed by the same individual or by different individuals of the opposite sex.

This process is usually slow and complex compared to asexual reproduction. The organisms so produced are generically diverse. Thus they can evolve along with the changing climatic conditions. Humans and many multicellular organisms exhibit a sexual mode of reproduction.

Reproduction in plants

Plants reproduces by sexual and asexual means vegetative reproduction is the main mode of plant reproduction. Roots such as a corm, stem tuber, shizomes and stolon undergo vegetative propagation.

Sexual reproduction in plants take place through pollination in which the pollen grains from the another of a male flower transfer to the stigma of the female flower.

Sexual Reproduction in plants

A few plants produce seeds without fertilization and the process is called apomixes. Here the ovule or the ovary gives rise to new seeds.

Reproduction in Animals

Animals reproduce sexually as well as asexually. Sexual reproduction involves the fusion of male and female gametes. This process is known as fertilization. Fertilization can be external or internal. External fertilization is the process in which the male sperm fertilizes the female egg outside the female's body. On the contrary, in internal fertilization, the fusion of male and female gametes takes place inside the body of female.

A sexual reproduction involves reproduction processes such as binary fission, budding fragmentation etc. The organisms have no reproductive system and therefore no formation of male and female gametes takes place.

Specifically the process by which plants and animals give rise to offspring and which fundamentally consists of the segregation of a portion of the parental body by a sexual or an asexual process and its subsequent growth and differentiation into a new individual

2- Something reproduced: copy

3- Young seeding trees in a forest reproduction, Duplicate, copy, fascinile, Replica- means a thing made to closely resemble another.

4- Reproduction-implies an exact or close imitation of an existing thing.

5- Duplicate-implies a double or counterpart exactly corresponding to another thing.

6- Copy- applies especially to one of number of things reproduced mechanically

7- FACSIMILE- Suggests a close reproduction often of graphic matter that may differ in scale.

8- REPLICA- Inplies the exact reproduction of particular item in all details. But not always in the same scale.

Reproduction (or procreation or breeding) is the biological process by which a new individual organism- offspring-are produced from their parent or parents. Reproduction is a fundamental feature of all known life; each individual organism exists as the result of reproduction. This is a universal process of all living things is present on this universe. There are two forms of reproduction: Asexual and sexual- These are universal processes of forming a new individual species.

In asexual reproduction, an organism can reproduce without the involvement of another organism. Asexual reproduction is not limited to single-called organism. The cloning of an organism is a form of asexual reproduction. By asexual reproduction, an organism creates a genetically similar or individual copy of itself.

The evolution of sexual reproduction is a major puzzle for biologists. The two fold cost of sexual reproduction is that only 50% of organisms reproduce and organisms only pass on 50% of the genes.

Sexual reproduction typically requires the sexual interaction of two specialized reproductive cells, called gametes, which contain half the number of chromosomes of normal cells and are created by meiosis, with typically a male fertilizing a female of the same species to create a fertilized zygote. This produces offspring organisms whose genetic characteristics are derived from those of the two particular organisms.

www.ingramcontent.com/pod-product-compliance
Ingram Content Group UK Ltd.
Pitfield, Milton Keynes, MK11 3LW, UK
UKHW062000290726
14090UKWH00021B/1304